# Survey Weights: A Step-by-Step Guide to Calculation

# Survey Weights: A Step-by-Step Guide to Calculation

RICHARD VALLIANT
*Universities of Michigan & Maryland*

JILL A. DEVER
*RTI International (Washington, DC)*

A Stata Press Publication
StataCorp LLC
College Station, Texas

Copyright © 2018 StataCorp LLC
All rights reserved. First edition 2018

Published by Stata Press, 4905 Lakeway Drive, College Station, Texas 77845
Typeset in LaTeX $2_\varepsilon$
Printed in the United States of America

10  9  8  7  6  5  4  3  2  1

Print ISBN-10: 1-59718-260-5
Print ISBN-13: 978-1-59718-260-7
ePub ISBN-10: 1-59718-261-3
ePub ISBN-13: 978-1-59718-261-4
Mobi ISBN-10: 1-59718-262-1
Mobi ISBN-13: 978-1-59718-262-1

Library of Congress Control Number: 2017960405

# Acknowledgments

We are indebted to several people who have answered questions and encouraged us in the writing of this book. Jeff Pitblado of StataCorp programmed `svycal`, which is a new Stata procedure that can handle raking, poststratification, general regression, and more general calibration estimation. He also answered many specific Stata questions. This book would not have been possible without him.

Matthias Schonlau at the University of Waterloo provided valuable assistance on how to use his `boost` plug-in and how to tune parameters in boosting. Nicholas Winter helped us several times with questions about his `survwgt` package, which seems to get far less publicity than it deserves. Stas Kolenikov advised us on Stata's general capabilities and on his `ipfraking` raking procedure, which is also a useful tool for computing survey weights.

We thank Frauke Kreuter for many things. Her boundless energy and endless fount of ideas have pushed us along for years. Finally, we thank our spouses, Carla Maffeo and Vince Iannacchione, for their support throughout this several-year project.

Richard Valliant
Jill A. Dever

November 2017

# Contents

# Figures

# Preface

Many data analysts use survey data and understand the general purpose of survey weights. However, they may not have studied the details of how weights are computed, nor do they understand the purpose of different steps used in weighting. *Survey Weights: A Step-by-step Guide to Calculation* is intended to fill these gaps in understanding. Throughout the book, we explain the theoretical rationale for why steps are done. Plus, we include many examples that give analysts tools for actually computing weights themselves in Stata.

We assume that the reader is familiar with Stata. If not, Kohler and Kreuter (2012) provide a good introduction.

Finally, we also assume that the reader has some applied sampling experience and knowledge of "lite" theory. Concepts of with-replacement versus without-replacement sampling and single- versus multistage designs should be familiar. Sources for sampling theory and associated applications abound, including Valliant, Dever, and Kreuter (2013), Lohr (2010), and Särndal, Swensson, and Wretman (1992), to name just a few.

## Structure of the book

When faced with a new dataset, it is good practice to ask yourself a few questions before analyzing the data. For example,

- Am I dealing with a sample, or does the dataset contain a whole population?
- If it is a sample, how was it selected?
- What is my goal for the analysis? Am I trying to draw inference to the population?
- Do I need to weight my sample to project it to the population?
- Do I need to weight my data to compensate for the fact that the sample does not correctly cover the desired population?

Some datasets you encounter might already contain weights, and it is useful to understand how they were constructed. If you collect data yourself, you might need to construct weights on your own. In both cases, this book will give useful guidance, both for the construction and for the use of survey weights. This book can be read straight through but can also serve as a reference for specific procedures you may need to understand. You can skip around to particular topics and look at the examples for useful code.

We start our book with a general introduction to survey weighting in chapter 1. Weights are intended to project a sample to some larger population. The steps in weight calculation can be justified in different ways, depending on whether a probability or nonprobability sample is used. An overview of the typical steps is given in this chapter, including a flowchart of the steps.

Chapter 2 covers the initial weighting steps in probability samples. The first step is to compute base weights calculated as the inverse of selection probabilities. In some applications, because of inadequate information, it is unclear whether some sample units are actually eligible for the survey. Adjustments can be made to the known eligible units to account for those with an unknown status.

Most surveys suffer from some degree of nonresponse. Chapter 3 reviews methods of nonresponse adjustment. A typical approach is to put sample units into groups (cells) based on characteristics of the units or estimates of the probabilities that units respond to the survey. This chapter also covers another option for cell creation—using machine learning algorithms like CART, random forests, or boosting to classify units.

Chapter 4 covers calibration or adjusting weights so that sample estimates of totals for a set of variables equal their corresponding population totals. Calibration is an important step in correcting coverage problems and nonresponse and, in addition, can also reduce variances.

Chapter 5 discusses options for variance estimation, including exact formulas, linearization, and replication. Using multiple adjustments in weight calculation, as described in the previous chapters, does affect the variance of point estimates of descriptive quantities like means and totals. We illustrate how these multiple effects can be reflected using replication variances.

Not all sets of survey data are selected via probability samples. Even if the initial sample is probability, an investigator often loses control over which units actually provide data. This is especially true in the current climate, in which people, businesses, and institutions are progressively becoming more resistant to cooperating. Chapter 6 describes methods to weight nonprobability samples. The general thinking about estimating propensities of cooperation and using calibration models, covered in chapters 3 and 4, can be adapted to the nonprobability situation.

Chapter 7 covers a few special situations. Normalized weights are scaled so that they sum to the number of units in the sample—not to an estimate of the population size. Although we do not recommend them, normalized weights are used in some applications, particularly in public opinion surveys. Other topics in this chapter include datasets with multiple weights, two-phase sampling, and weights for composite estimation. Some survey datasets come with more than one weight for each case, especially when subsamples of units are selected for different purposes. Two-phase sampling is often used when more intensive efforts are made to convert nonrespondents for a subsample of cases. Composite weighting is used to combine different samples from different frames such as persons with landline telephones and persons with cell phones. This chapter also covers whether to use survey weights when fitting models. We describe the issues that need to be considered and give some analyses that can be done when deciding whether to use weights in fitting linear and nonlinear models from survey data.

Chapter 8 covers the unexciting but essential procedures needed for quality control when computing survey weights. An orderly system needs to be laid out in advance to guide the sequence of weighting steps, to list quality checks that will be made at every step, and to document the entire process.

# Data files and programs for this book

The data and program files used in the examples are available on the Internet. You can access these files from within Stata or by downloading a zip archive. For either method, we suggest that you create a new directory and download the materials there.

- If the machine you are using to run Stata is connected to the Internet, you can download the files from within Stata. To do this, type the following commands in the Stata Command window:

  ```
  . net from http://www.stata-press.com/data/svywt/
  . net describe svywt
  . net install svywt
  . net get svywt
  ```

  Notice that the statements above are prefaced by "." as in the Stata Results window. We use this convention throughout the book.

- The files are also stored as a zip archive, which you can download by pointing your browser to http://www.stata-press.com/data/svywt/svywt.zip.

  To extract the file `svywt.zip`, create a new folder, for example, `svywt`, copy `svywt.zip` into this folder, and unzip the file `svywt.zip` using any program that can extract zip archives. Make sure to preserve the subdirectory structure contained in the zip file.

Throughout the book, we assume that your current working directory (folder) is the directory where you have stored our files. This is important if you want to reproduce our examples.

Ensure that you do not replace our files with a modified version of the same file; avoid using the command `save, replace` while working with our files.

# Glossary of acronyms

| | |
|---|---|
| BRR | balanced repeated replication |
| cv | coefficient of variation |
| deff | design effect |
| ENR | eligible nonrespondents |
| epsem | equal probability sampling and estimation method |
| ER | eligible respondents |
| fpc | finite population correction |
| GREG | general regression |
| IN | ineligible |
| KN | known eligibility |
| MAR | missing at random |
| MCAR | missing completely at random |
| mos | measure of size |
| NMAR | not missing at random |
| NR | nonresponse |
| OLS | ordinary least square |
| pps | probability proportional to size |
| PSU | primary sampling unit |
| pwr | probability with replacement |
| relvar | relative variance (square of cv) |
| SE | standard error |
| srs | simple random sampling |
| srswor | simple random sampling without replacement |
| srswr | simple random sampling with replacement |
| stsrs | stratified simple random sample |
| stsrswor | stratified simple random sample without replacement |

| | |
|---|---|
| UNK | unknown eligibility |
| UWE | unequal weighting effect |
| VarStrat | variance strata |
| VarUnit | variance unit |

# 1 Overview of weighting

Survey datasets, released to the public through a general- or restricted-use agreement, usually come with at least one analysis weight for each respondent record in the sample. (In some applications, more than one weight may be provided for each record for special purposes discussed in chapter 7.) Analysts interested in calculating population estimates are told to use the same set of weights for all analyses—means, totals, linear and nonlinear models, etc. The benefits and drawbacks of a single analysis weight compared with multiple weights for tailored analytic objectives is reviewed in section 1.3.

Analysis weights are designed to

1. account for the probabilities used to select units (in cases where random sampling is used);

2. adjust in cases where it cannot be determined whether some sample units are members of the population under study;

3. adjust for eligible units that do not respond to the survey to limit the effects of nonresponse bias; and

4. incorporate external data to reduce standard errors of estimates and to compensate when the sample does not correctly cover the desired population.

However, unless you are the developer of the weights, the datasets typically contain the final analysis weights and not the adjustments for the above conditions.

Survey statisticians usually think of weighting in the context of probability samples, where units are selected by some random means from a well-defined population. All four steps above can be applied to probability samples. However, because of the current popularity of volunteer web panels and other kinds of "found" data, how to weight nonprobability samples is also worth considering. For those samples, steps 3 and 4 can be used (see chapter 6).

This chapter gives an overview of the purposes of weighting, underlying theory and sampling methods, and some problems that are considered when constructing a set of weights. The information in this chapter forms the basis for our discussion in this book. Specifically, the last section of this chapter contains an overview of weighting procedures and serves as an important reference for the remaining chapters.

## 1.1  Reasons for weighting

The fundamental reason for using weights when analyzing survey data is to produce estimates for some larger target population, that is, population inference. Ideally, the estimates will a) be unbiased or consistent in a sense described later, b) have standard errors that are as small as is feasible given the sample size and sample design, and c) correct for deficiencies in how the sample covers the desired population. Depending on the type of analysis being done, the population may be some well-defined finite population, like all adults aged 18 years and older in a country. The goal when making other estimates, like those of parameters in a regression model, may be to represent some population that, at least conceptually, is broader than any given finite population.

A finite population is a collection of units (also referred to as elements or cases) that could, in principle, be completely listed so that a census could be conducted to collect data from each unit. Examples, in addition to the adult population mentioned above, are elementary schools in a county, hospitals in a state, registered voters in a city, and retail business establishments in a province.

Defining the units that are members of a finite population (that is, eligible units) may require some thought, depending on the type of population. Whether a person is age 18 or older (and eligible to vote in the United States) seems straightforward, but defining what constitutes a business establishment is more difficult. Often, the composition of a population can change over time so that a specific time period must be part of the definition of the population. For example, a finite population of registered voters might be defined as those persons who are registered as of the date an election is to be held. The January labor force in a country may be defined as all persons who are employed or unemployed but seeking a job during the second week of that month.

### Target populations and sampling frames

Understanding the distinction between a target population (also referred to as the universe of all population members or just universe) and a sampling frame is important when assessing the strengths and weaknesses of a sample. The target population is the population for which inferences or estimates are desired. The sampling frame is the set of units from which the sample is selected. Ideally, the sampling frame and the target population are the same. In that case, we say that the sampling frame completely covers the target population. However, there are many instances where the two do not coincide.

Figure 1.1 is a diagram of how the universe $U$, the sampling frame $F$, the sample $s$, and the complement of the sample within $U$, $s^c$, might be related. The frame $F$ can omit some eligible units (undercoverage) and include other ineligible units (overcoverage). The eligibles in the frame in figure 1.1 are denoted by the intersection of $U$ and $F$, $U \cap F$, while the ineligibles in the frame are denoted by those not included in $U$, $F - U$. The sample $s$ can include both eligible units in $s \cap U$ and ineligible units in $s \cap (F - U)$. The latter condition occurs if the true eligibility of the units on the frame is unknown when the sample is selected. In the figure, the eligible units that are not in the frame or sample are denoted by $U - F$. In the ideal situation, the frame completely covers the population so that $F = U$. The purpose of weights is to project the eligible sample, $s \cap U$, to the full universe, $U$. As is apparent from the figure, this will require eliminating the ineligible units from the sample (or at least those known to be ineligible) if such information is not available to remove them initially from the frame. We also hope to use the sample to represent the units in the universe that were not in the frame, $U - F$, and consequently had no chance of being selected for the sample. One of the functions of weighting is to attempt to correct for such coverage errors.

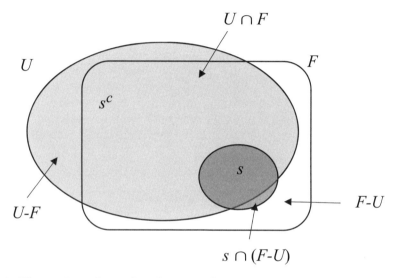

Figure 1.1. Illustration of sampling frame with over- and undercoverage of target population

The most straightforward case of a sampling frame is a list of every unit in the target population. For example, if we want to survey the members of some professional organization like the Royal Statistical Society (target population), a current membership list (sampling frame) may be available from which the sample can be selected. However, if the list was somewhat outdated because it omits people who became members in the last month, or it still contains some deceased members, the frame would have coverage errors. Current members not covered by the list cannot be sampled, although they would be eligible for the study. Past members covered by the list can be sampled, although they are ineligible for the study.

A complete list of the members of the target population is not always available, but it may be possible to construct a frame that does cover the whole population. For example, in household surveys, a list of all households or people who live in them is not available in many countries. Even if a government agency has such a list, it may not be accessible to private survey organizations. Standard practice is to compile a frame in stages. For example, a sample of geographic areas is selected, perhaps in several stages, and a list of households is compiled only within the sample areas. When executed properly, this technique will provide virtually complete coverage. However, in practice, achieving complete coverage of a household population is difficult or impossible. Even the Current Population Survey in the United States, which is quite well conducted, had about 15% undercoverage of persons in 2013 (U.S. Census Bureau 2013).

### Types of statistics

Descriptive statistics, like means or totals, are usually thought of as estimates of the quantities that would be obtained if a census were conducted of a finite population. For example, if the estimate is for the mean salary and wage income per person in a particular calendar year, the target for the sample estimate is the mean that would be obtained if all persons in the finite population were enumerated and the income collected for each. A population total is another example of a descriptive statistic. The finite population total itself is $t_y = \sum_{i \in U} y_i$, where $U$ is the set of all $N$ units in the population. Suppose a sample of $n$ units is selected from the population. An estimated total often has the form $\widehat{t}_y = \sum_{i \in s} w_i y_i$, where $i$ denotes a unit, $s$ is the set of $n$ units in the sample, $w_i$ is a weight assigned to unit $i$, and $y_i$ is the value of a data item collected for unit $i$. Weights that are appropriate for estimating totals are generally larger than or equal to 1 because $n < N$, and the weights need to inflate the sample to the larger population. In fact, for $y_i = 1$ for all units in the sample, $\sum_{i \in s} w_i = \widehat{N}$, is an estimate of the finite population size. Note that we use "hat notation" to signify estimates such as $\widehat{N}$, the estimate of the true population size, $N$.

Survey weights can also be used to estimate more complicated quantities like model parameters. For example, consider the simple linear regression model $y_i = \alpha + \beta x_i + \varepsilon_i$, where $\alpha$ and $\beta$ are parameters, and the $\varepsilon_i$'s are errors that are independent under the model with mean 0 and variance $\sigma^2$. The survey-weighted estimate of the slope computed by Stata and other software that handle survey data is

$$\widehat{\beta} = \frac{\sum_{i \in s} w_i \left(y_i - \overline{y}_w\right)\left(x_i - \overline{x}_w\right)}{\sum_{i \in s} w_i (x_i - \overline{x}_w)^2} = \frac{\sum_{i \in s} w_i x_i y_i - \left(\sum_{i \in s} w_i\right)\overline{x}_w \overline{y}_w}{\sum_{i \in s} w_i x_i^2 - \left(\sum_{i \in s} w_i\right)\overline{x}_w^2}$$

with $\overline{y}_w = \sum_{i \in s} w_i y_i / \sum_{i \in s} w_i$ and $\overline{x}_w$ defined similarly. As the second expression for $\widehat{\beta}$ shows, the estimated slope is a combination of several different estimated totals. Thus, estimated totals are frequently the building blocks for calculating quantities that are more complicated.

Estimates of model parameters can be interpreted in one of two ways. The first is the same as for descriptive statistics: $\widehat{\beta}$ estimates the value that would be obtained if a census were done and the model fit via ordinary least squares (that is, without weights) for the full, finite population. The second interpretation is, perhaps, more subtle: $\widehat{\beta}$ estimates a model parameter that applies to units beyond those in the fixed, finite population from which the sample was drawn. For example, suppose a sample of persons is selected in April 2015, and an analyst regresses personal income on years of education. The analyst is probably interested in making a statement about the effect of education on income not just in April 2015 but also without regard to the month when the survey happened to have been done. This also raises the question of whether the survey weights should be used at all in model fitting—a topic we address in more detail in chapter 7.

## 1.2   Probability sampling versus nonprobability sampling

Survey samples can be selected in one of two ways. The first is through a defined probabilistic method that is reproducible and is labeled as probability sampling. The second is by way of an undefined sampling mechanism that is not exactly reproducible, known in the survey world most recently as nonprobability sampling. The method that is used affects how weights are calculated.

Probability sampling means that units are selected from the finite population in some random manner. Probability sampling has a very specific, technical definition given in Särndal, Swensson, and Wretman (1992) and other books on sampling theory. Four conditions must be satisfied for a sample to be a probability sample:

1. The set of all samples that are possible to obtain with the specified sampling procedure can (in principle) be enumerated.

2. Each possible sample $s$ has a known probability of selection, $p(s)$.

3. Every unit in the target population has a knowable, nonzero probability of selection.

4. One set of sample units is selected with the probability associated with the set.

If a probability sample is selected, the first step in weighting is to compute an initial or base weight for each unit, which is the inverse of its selection probability. Base weights are mentioned in section 1.7 and described further in chapter 2.

Although the requirements above seem to imply that every possible sample would have to be identified, a probability sample can be selected in a way that does not require listing all the possibilities. Standard procedures also require only that the probabilities of selection of individual units be tracked—values of $p(s)$ are unnecessary.

Probability samples are the standard for governmental surveys that publish official statistics, like the unemployment rate, the inflation rate, and statistics on the health of a population. If time and budget allow, other surveys like pre-election polls may also select probability samples. This method of sampling provides one mathematical basis for making estimates, as discussed in section 1.3. It also adds a degree of face validity to the results. A survey designer cannot be accused of injecting conscious or unconscious biases into the selection of units when a random mechanism is used to decide which units are picked for the sample. Because every element in the population has a chance of being selected for a sample, the sample covers the entire population. If enough information is available on the frame in advance of sampling, a survey designer can also control the distribution of the sample among various subgroups.

On the other hand, it may be cheaper and quicker, or only feasible, to acquire sample cases without a defined probability method (that is, by using nonprobability methods). Characteristics of interest may be time sensitive, and sampling may have to be done in the field by data collectors. Asking visitors to a website to participate in a survey voluntarily is one way that is currently being used to collect sample data. For example, a survey sponsor can inexpensively accumulate a huge number of persons this way and request that they become part of a panel that will cooperate in future surveys. One obvious criticism of this approach is that only a selective group of persons may visit the website used for recruiting. The persons who volunteer may be a poor cross-section of the population at large; that is, the sample may be subject to severe coverage error. Of course, this sort of criticism can be levied against any sample where there is no control or limited control over which sample units actually participate. A committee of the American Association for Public Opinion Research (AAPOR) conducted an extensive review of nonprobability samples (Baker et al. 2013b). Elliott and Valliant (2017) review the theoretical issues with inference from nonprobability samples and some of the methods that have been proposed for estimation. We investigate weighting for nonprobability surveys in detail in chapter 6.

Samples often live in some fuzzy netherworld between probability and nonprobability. A sample may begin as a probability sample but then suffer from a high rate of nonresponse. Because the survey designer cannot completely control which units respond, the set of units that ultimately respond may not reflect the intended probability sample. Nevertheless, starting with a probability sample selected from a high-quality frame provides some degree of comfort that a sample will have limited coverage errors.

A web panel of persons is a case in point. One approach to forming a web panel is to select a large telephone sample of households and request the cooperation of all persons over a certain age. The initial sample may be a probability sample of all telephone numbers known to be in use, but the resulting panel can suffer from at least two problems. If any phone numbers are omitted from the sampling frame, an undercoverage problem may result if the omitted portion differs from those on the frame. For example, if a frame uses only landline phones, then households with only cell phones cannot be selected. Telephone surveys also often have poor response rates—30% or less is common in the United States. If the respondents are not randomly spread over the initial sample, then there may be nonresponse bias, another source of potential undercoverage.

As discussed in chapters 3 and 4, weights can be constructed that attempt to adjust for both coverage and nonresponse error. The success of these adjustments depends on strong assumptions that are described there.

## 1.3 Theories of population inference

Weights and estimators are intimately linked because, as noted in section 1.1, many statistics are constructed as combinations of estimated totals that have the form $\hat{t}_y = \sum_{i \in s} w_i y_i$. Consequently, a goal in creating weights is to construct (approximately) unbiased and efficient estimators. To define terms like unbiased and efficient, statistical theory is needed. The three approaches used to analyze properties of estimators in survey sampling are

1. design based, which is also called repeated sampling or randomization based;

2. model based; and

3. model assisted.

Like other parts of statistics, the theoretical distribution of an estimator is used to identify its properties in sampling theory. For example, is the distribution centered on the population value to be estimated? Is the distribution concentrated around that true value, or is it spread widely?

In the design-based approach, the distribution of an estimator is generated by thinking of the values that this estimator could have in each of the samples that could be selected using a particular sampling plan (that is, repeated sampling). In the model-based approach, the $y$ values are treated as being realizations from some mathematical model (see, for example, Valliant, Dorfman, and Royall [2000]). A distribution of an estimator is then the set of values that an estimator could take on under the model, given the particular sample that was selected. Model-based inference is particularly relevant for nonprobability samples, discussed in chapter 6. In the model-assisted approach, a model is used to help form an efficient estimator, but the properties of the estimator are analyzed with respect to repeated sampling.

An estimator is unbiased in repeated sampling or "design unbiased" if the average value of the estimates across all the possible samples that could be selected under a particular sample design equals the finite population value of whatever is being estimated. This says that $E_I(\hat{\theta}) = \theta$ where $E_I$ is the "expectation" (average) with respect to the sampling design, and $\hat{\theta}$ is the estimated value for some population quantity $\theta$ like a mean or a total.

An estimator is "model unbiased" if the difference between the value of an estimator and the population value is zero when the difference is averaged over the values that could be generated under the model. That is, $E_M(\hat{\theta} - \theta) = 0$ where $E_M$ is the expectation with respect to the model.

A more important, but somewhat more theoretical property, is "consistency", which can be defined for either the design- or model-based approach. Roughly speaking, an estimator is said to be consistent if it gets closer and closer to the value it is supposed to be estimating as the sample size increases.

There are pros and cons with each of these approaches. The design-based approach is model-free in the sense that statistical properties do not depend on some assumed population model being true. One set of weights can be constructed that will have good design-based properties and be used for all estimates. This is a major practical advantage when preparing datasets for analysts who are not specialists in sampling theory. However, the design-based approach does not provide any systematic guidance on how to construct estimators and their implied weights. Another criticism is that design-based properties, like repeated-sampling unbiasedness, do involve averaging over samples that may be much different from the one that was actually selected. Thus, having a property like design unbiasedness does not tell you whether the estimate from a particular sample is likely to be close to the target value.

A pro for the model-based approach is that it does provide guidance on how best to construct an estimator. For example, if a $y$ depends on a covariate $x$, that relationship can be exploited, as in a regular regression problem, to construct a better estimator of a population mean or total than the weighted sample mean that uses only inverse selection probabilities as weights (see discussion of base weights in section 1.7 and chapter 2). Another pro for the model-based approach is that it does compute properties for the particular sample that was selected rather than averaging over all possible samples. On the other hand, if the model used for finding properties is wrong, then inferences about population values may be wrong. Another con is that the same model will not necessarily hold for all variables in a multipurpose survey, which means that the same estimator (and resulting set of weights) will not be equally efficient for all $y$'s.

The model-assisted approach is a compromise between the design- and model-based approaches in which models are used to construct estimators, but the repeated sampling distribution is used for inference. This approach is probably closest to the way practitioners think about the problem of estimation and weight construction. Using the model-assisted technique, one can construct estimators and weights that have good design-based properties for all $y$'s in a survey and reasonably good model-based properties for some of the $y$'s. However, a single set of weights will not be model-efficient for all types of estimates. For example, by using a linear model with a particular set of covariates to construct weights, low variance estimates of totals will be produced for $y$'s that follow that model, but for $y$'s that follow a nonlinear model, the estimated totals may not be efficient at all.

One approach that we have not mentioned is the Bayesian approach, which seems to be getting more attention in sampling and other areas of statistics. Bayesian inference is an extension of model-based inference. Additional model distributions are assumed to hold for the parameters in a model. For example, in the model $y_i = \alpha + \beta x_i + \varepsilon_i$, the parameters $\alpha$ and $\beta$ are treated as random and having some distribution like normal. The variance of the error term may also be assigned a distribution. Bayesian theory for finite population estimation was introduced in Ericson (1969); many results are summarized in Ghosh and Meeden (1997) and Ghosh (2009). Like the model-based approach, Bayesian methods are good ways of generating efficient estimators. Bayes' theorem is used to compute posterior distributions of parameters that are used in estimating means, totals, and other quantities. As a result, inferences are conditional on both the set of sample units that was selected and the $y$ values for those units. The objection about averaging over data that we did not actually see is removed. As with a non-Bayesian model-based approach, objections are that every $y$ variable may require its own estimation procedure, the model assumptions may be wrong, and a single set of weights cannot be produced for use with all estimators. In some cases, weights do not flow out of a Bayes procedure at all.

Although the Bayesian approach has some strong advocates (for example, Little [2004]), it is currently used in large-scale surveys only in some special applications like small area estimation. The probability sampling techniques we cover in this book are non-Bayesian (although they may, in some cases, have a Bayesian interpretation). We briefly discuss a type of Bayesian estimation for nonprobability samples in chapter 6.

# 1.4   Techniques used in probability sampling

Probability samples are selected through several methods that are geared toward improving the precision of estimators, facilitating fieldwork, and keeping costs under control. The particular sampling scheme used to select a sample dictates the structure of the initial weights that may be adjusted to limit bias or improve precision. Thus, we list some of the main techniques below. Many books on theoretical and applied sampling, for example, Cochran (1977), Levy and Lemeshow (2008), Lohr (2010), Särndal, Swensson, and Wretman (1992), and Valliant, Dever, and Kreuter (2013), give details that we only sketch here. One way of categorizing probability samples is by the method used for random sampling, whether the survey uses stratification or clustering, and by how many stages of sampling are used. We discuss each of these below.

### Methods of random sampling

The simplest technique is equal probability sampling in which each unit in the population has the same selection probability. This is sometimes known as equal probability sampling and estimation method (epsem) or self-weighting (Kish 1965). An epsem sample can be selected via simple random sampling (srs), either with or without replacement from a sampling frame such as a membership roster of an organization. Another way of selecting an epsem sample is systematic sampling in which a list is sorted in some order, a random starting place is selected, and the sample is selected by skipping systematically down the list. For example, field interviewers may be instructed to interview every fifth house on a defined path within a randomly chosen neighborhood. An epsem sample can also be selected in several stages as noted below.

Probability proportionate to size (pps) (or, more generally, sampling with unequal probabilities) is a method of sampling units with different probabilities depending on their relative sizes. For example, hospitals might be sampled with probabilities proportional to their numbers of inpatient beds. In a household survey, geographic areas may be selected with probabilities proportional to their population counts. If the measure of size used for sampling is related to the items that will be collected, pps sampling can be extremely efficient. These samples can be selected in various ways, including systematic.

Two methods that may be used in special applications are Bernoulli and Poisson sampling. In Bernoulli sampling, each unit in a population is given the same independent chance $\pi$ of selection. The chance of selection is the same for every unit; consequently, this is epsem. Poisson sampling differs only in that each unit can have a different selection probability, $\pi_i$. These methods are useful when the units in a population become available only over an extended period of time. An example is the population of tax returns filed with a governmental agency. Filings by taxpayers usually occur over a range of months. Bernoulli or Poisson sampling allows a sample to be selected as the returns flow in rather than waiting until the end of the tax filing season when the full population is in hand.

### Stratification

A population is divided into mutually exclusive groups, or "strata", that collectively cover the entire population. A sample is then selected independently from each of the groups. Stratification can be used to 1) avoid selecting a sample that is poorly distributed across the population, as could occur in srs; 2) assure that important subgroups are represented in the sample or possibly overrepresented to boost power for some analytic objective; 3) form administrative groups, for example, ones where different data collection methods might be used; 4) manage the budget by accounting for cost differentials among strata; and 5) reduce variances by using an efficient allocation of the sample to strata.

### Clustering

Units are assigned to groups or clusters, and a sample of the groups is selected. This technique is often used for cost-control purposes to reduce the number of locations where data must be collected or in cases where a complete list of all population units is not available in advance of sampling. A list of population units must be compiled within only the sampled clusters. Three examples are schools, which can be considered as clusters of students; counties, which are clusters of households; and business establishments, which are clusters of employees.

### Stages of sampling

Single-stage sampling is used when a complete frame is available and the method of data collection does not require that sample units be geographically or administratively clustered to facilitate data collection. For example, area household surveys may require personal visits by data collectors to obtain responses and biological measurements. Military surveys may require additional clustering by administrative unit to gain permission prior to recruiting a subsample of soldiers within the unit. Conversely, telephone surveys of households are usually single stage if a reasonably complete list of phone numbers is available. Some surveys of establishments are single stage if data can be collected by phone, mail, or electronic medium (for example, email invitation for a self-administered, web-based questionnaire).

Samples are sometimes selected in multiple stages, either as a way of reducing costs or because there is no feasible alternative. For example, a sample of households may be obtained by sampling counties or groups of counties at the first stage, census blocks at the second stage, and households at the third stage. When data collection requires personal interviews, sampling in this way reduces travel costs by clustering sample households geographically. It also allows current lists of households to be compiled in the field if a complete list of households and their addresses is not available from some administrative source. Stratification, clustering, and unequal probability sampling are all typically used in multistage sampling.

The method of random selection, the use of stratification and clustering, and the number of stages in sampling may all need to be considered when computing base weights—the inverse of the selection probabilities. Base weights are affected when strata have different sampling rates. For multistage sampling, the selection probability of each unit at each stage of sampling must be tracked and ideally stored in a master database (see chapter 8). In short, any design feature that affects selection probabilities should be considered when computing weights. The four features above also affect how variances and standard errors should be estimated, as discussed in chapter 5.

## 1.5   Weighting versus imputation

Weights are used to project information obtained from a sample (or a portion of the sample if not all eligible sample members participate) to the target population. This requires using the eligible responding sample units, $s_{\mathrm{ER}}$ ($s \cap U$ in the sample $s$ in figure 1.1) to project values for the eligible units in the target population that are not in the responding sample, $s_{\mathrm{ER}}^c \cup (U - F)$. This is a form of missing-data problem—values for the units in $s_{\mathrm{ER}}$ are observed, but units in $s_{\mathrm{ER}}^c$ (eligible units on the frame but not in the responding sample) and $U - F$ (eligible units not on the frame) are missing. In an estimator like $\widehat{t}_y = \sum_{i \in s_{\mathrm{ER}}} w_i y_i$, the usual intuitive description of the weight $w_i$ is that unit $i$ represents itself plus $w_i - 1$ others. One way to think of this is that the value $y_i$ is imputed to $w_i - 1$ other units. Another way of writing the estimator of a total, $\widehat{t}_y = \sum_{i \in s_{\mathrm{ER}}} w_i y_i$, is

$$
\begin{aligned}
\widehat{t}_y &= \sum_{i \in s_{\mathrm{ER}}} y_i + \sum_{i \in s_{\mathrm{ER}}} (w_i - 1) \, y_i \\
&= t_{s_{\mathrm{ER}}} + \widehat{t}_{U - s_{\mathrm{ER}}}
\end{aligned}
$$

where $t_{s_{\text{ER}}} = \sum_{i \in s_{\text{ER}}} y_i$ is the sample sum, and $\widehat{t}_{U-s_{\text{ER}}} = \sum_{i \in s_{\text{ER}}} (w_i - 1) y_i$ is a prediction of the nonsample sum in the population, $t_{U-s_{\text{ER}}} = \sum_{U-s_{\text{ER}}} y_i$. That is, $\sum_{i \in s_{\text{ER}}} (w_i - 1) y_i$ functions as a predictor of $\sum_{i \in U-s_{\text{ER}}} y_i$, which is another way of saying that $\widehat{t}_y$ contains an implied imputation for $t_{U-s_{\text{ER}}}$.

Many methods are available to impute for $t_{U-s_{\text{ER}}}$. Hot deck, regression, and nearest neighborhood are some of the possibilities. Kim and Shao (2014) cover many of the options. Mass imputation is when individual unit-level imputations are made for all variables in the analysis dataset (Kovar and Whitridge 1995). Because the number of units in $U - s_{\text{ER}}$ is typically large, weighting is the standard procedure in sample surveys rather than mass imputation for the nonsample units. Because of the wealth of existing information and the focus of this book, we leave the discussion of imputation to the references above and other related citations.

## 1.6   Disposition codes

Numeric codes that describe the current or final data collection status of each sample unit are known as disposition codes (Valliant, Dever, and Kreuter 2013, chap. 6). The AAPOR document, *Standard Definitions: Final Dispositions of Case Codes and Outcome Rates for Surveys* (AAPOR 2016) provides a list of recommended disposition codes. The AAPOR report lists codes that can be used for telephone and in-person household surveys, mail surveys, and Internet surveys. The AAPOR list is elaborate but can be mapped into the following four groups, which are useful for computing weights:

1. Eligible cases for which a sufficient amount of data are collected for use in analyses (eligible respondents, ER);

2. Eligible cases for which no data are collected (eligible nonrespondents, ENR);

3. Cases with unknown eligibility (UNK); and

4. Cases that are not eligible members of the target population (ineligible, IN).

We will also denote the set of cases whose eligibility is known (ER, ENR, and IN) as KN.

The codes are generally specific to each data collection agency, population being surveyed, and mode of the survey. As an example, table 1.1 shows the sample disposition codes recorded for the May 2004 Status of Forces Survey of Reserve Component Members (SOFReserves), a mail survey conducted by Defense Manpower Data Center (Defense Manpower Data Center 2004) of military reservists. A survey of households or establishments will likely have a different set of disposition codes.

Table 1.1. Terminology: Sample dispositions for the May 2004 SOFReserves study

| Disposition code | Description |
| --- | --- |
| 1 | Ineligible — based on check of updated personnel records |
| 2 | Ineligible — self/proxy report, deceased, ill, incarcerated, separated from military |
| 3 | Ineligible — survey self report |
| 4 | Complete eligible response |
| 5 | Incomplete eligible response |
| 8 | Refused — refusal, deployed, other refusal |
| 9 | Blank (returned questionnaire) |
| 10 | Postal nondelivery (PND) |
| 11 | Other nonrespondent |

Once data collection is finished, a final disposition code is assigned to each sample unit. Each code is mapped into the broad ER, ENR, UNK, and IN categories described above based on specifications ideally defined during the early stages of the study design and before data collection begins (chapter 8). These categories are then used in calculating adjustments to the base weights.

## 1.7  Flowchart of the weighting steps

As observed at the beginning of this chapter, computing weights for a probability sample involves several steps: base weights, adjustment for units whose eligibility for the survey cannot be determined, adjustment for nonresponse, and use of external data to improve estimators. Figure 1.2 is a flowchart showing the sequence of steps followed by many developers of analysis weights for surveys that begin with probability samples. Throughout the steps in figure 1.2, it is critical to set up a data processing system that allows each step to be done. This involves tracking the pieces of information for each record that are required for each step and, not incidentally, establishing quality controls to ensure that each step is done correctly. At each step of weight calculation, it is important to save the results for each record from that step to a central data file (see master database discussion in chapter 8).

### Step 1: Base weights

Base weights (inverse of selection probabilities) are calculated for every unit in the initial sample with respect to the sampling design and stages of selection. This even includes units that may later be dropped because they are ineligible, do not provide data, or are never released for data collection. All cases are retained after step 1 for subsequent processing. Note that, when units are selected without replacement, all base weights should have a value greater than or equal to one. We discuss additional quality assurance checks throughout the chapters.

### Step 2:  Unknown eligibility adjustment

In some surveys, there may be units whose eligibility cannot be determined—the unknowns (UNKs). For example, if the survey is to include persons whose age is 50 years or older, some people may refuse to disclose their age. If the survey uses in-person interviewing, some households cannot be contacted because no one is ever at home during the field period. As shown in the flowchart, when there are UNKs, the cases with known eligibility (KN = ER, ENR, and IN) have their weights adjusted. This usually consists of distributing the weights of the UNKs to the KNs, as described in section 2.2.

In step 2, the UNK and IN cases are removed and saved to separate files. Although it may be tempting to drop these cases entirely, the prudent approach is to save them for documentation and in case the weighting steps have to be redone for some reason. Also, the IN units may be used in a later weighting step (like step 4) if deemed appropriate. The eligible respondents and nonrespondents are then passed to step 3.

### Step 3:  Nonresponse adjustment

Respondents' weights are adjusted in this step to account for the ENRs. There are a variety of ways to do this, as covered in chapter 3. Cells may be formed based on covariate values known for ERs and ENRs. A response propensity model may be fit. A statistical classification algorithm, like a regression tree, may be used to put cases into bins. In each of the options, the weights of the ERs are increased to compensate for the fact that some eligible cases did not provide data.

The ENRs are saved to a separate file at the end of this step. The responding cases (and possibly INs) are then passed to the next step.

### Step 4:  Calibration

Statistics external to the survey are used in this step to either reduce variances or correct for coverage errors. This is termed "calibration" because the usual procedures result in certain estimated totals from the survey equaling some external reference totals. For example, weights may be calibrated in a household survey of persons so that the estimated total counts of persons in some age $\times$ race classes agree with the most recent census counts or demographic projections. (In market research, calibration is referred to as "sample balancing".)

There are several options for weight calibration, including poststratification, raking (that is, iterative proportional fitting), and general regression estimation. The external control totals may be population values, for example, census counts of persons or frame counts of beds in a hospital survey. Alternatively, they may be estimates from some other survey that is larger and better than your survey. Chapter 4 describes calibration in detail.

INs are included in this step along with the ERs only if the population controls (or estimates of them) are thought to also contain ineligibles. After calibration, the INs are removed from the analysis file.

**Step 5:  Analysis file**

The last step is simply to save the file of ERs with the final weights for each unit and their associated survey data.

Steps 1–4 may be implemented through multiple adjustments. For example, a survey of adolescents ages 12–17 in the United States typically requires parental permission prior to recruiting the adolescents into the study. Consequently, nonresponse can occur at two points in time—first for those without parental consent and second for those with parent consent but who subsequently refuse.

In your particular survey, some of or all the steps in figure 1.2 may be relevant. As shown in the flowchart, if a survey does not have cases of a particular type, then a step is bypassed. For example, if the eligibility of all sample cases is known, then step 2 can be skipped. This might be the case in a sample of hospitals where a complete frame is available, and the status of every sample hospital can be determined at the time of data collection. This may require some local knowledge if any hospitals have gone out of business since the frame was compiled. But, this kind of sleuthing is a routine part of fieldwork.

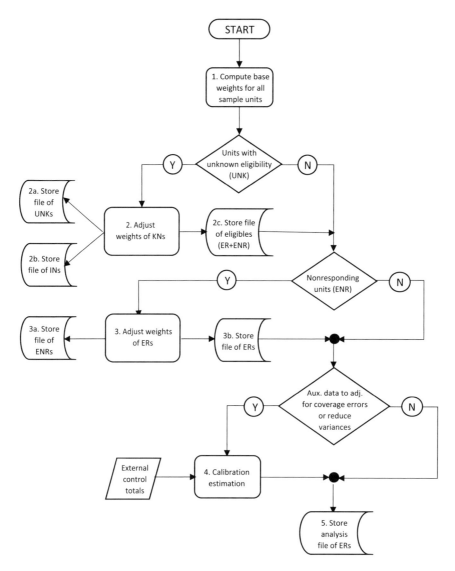

Figure 1.2.  Flowchart of steps used in weighting probability samples; KN = known eligibility, UNK = unknown eligibility, ER = eligible respondent, ENR = eligible nonrespondent, IN = ineligible

**Filenaming conventions.**

Each step in figure 1.2 will typically have a separate program associated with it. For example, we might have files named

```
1BaseWeights.do
2UNKadjustment.do
3NRadjustment.do
```

and so on. Using the step number as the first character of the file results in the programs sorting in the order that they should be executed in a file manager such as Windows File Explorer. Easily seeing this order is especially handy if you have to backtrack and re-do a step because disposition codes were updated or the input data changes for some reason. A name such as `1BaseWeights.do` also reminds you of the purpose of the program.

# 2 Initial steps in weighting probability samples

The first step in computing analysis weights for probability samples is to create base weights. As discussed in section 2.1, the base weights are calculated with respect to the sampling design. We provide base weights for some typical sampling designs along with Stata examples to support the discussion. Another source of example Stata code for selecting different types of samples and computing the appropriate base weights is the website of the Institute for Digital Research and Education at UCLA.[1]

The first decision point in the figure 1.2 flowchart is whether all sample units are members of the target population and hence eligible for the survey. If the eligibility status of some sample units is unknown, then we adjust the base weights for those units with a known eligibility status for those with unknown status (section 2.2).

## 2.1 Base weights

In without-replacement probability sampling, where a unit has only one chance of selection, base weights in a probability sample are calculated as the inverse of the selection probabilities. If $\pi_i$ is the probability of selection of unit $i$, its base weight is $d_{0i} = 1/\pi_i$. These will vary depending on the type of sample selected.

Another option is to select the sample with replacement, where units could be randomly chosen for the sample more than once. Although selecting a sample in this way would seldom be done in practice, with-replacement theory is very useful for developing variance estimators, as explained in chapter 5. In with-replacement sampling, the standard estimator is called a probability with replacement (pwr) estimator (see Särndal, Swensson, and Wretman [1992]). For the pwr estimator, the base weight is computed as the expected number of times a unit is selected, not the inverse of the selection probability as illustrated in the examples below.

**Simple random sampling**. In a simple random sample of size $n$ selected without replacement (srswor) from a frame of $N$ units, the selection probability of every sample unit is $n/N$, and the base weight is $N/n$. If the sample is selected with replacement (srswr), the selection probability of unit $i$ can be computed as one minus the probability that the unit is not selected in $n$ draws; that is, $\pi_i = 1 - (1 - 1/N)^n$. But, the pwr base weight would be computed as the inverse of $n$ times the probability that the unit

---

1. See https://stats.idre.ucla.edu/stata/seminars/svy-stata-intro.

is selected on a single draw. Because the single-draw probability for any unit is $1/N$, the pwr base weight is $N/n$ as in srswor.

In an srswor, the variance of estimators like a mean or total will contain a term called a "finite population correction factor" (fpc) equal to $1 - n/N$. We discuss this more in chapter 5, but illustrate calculations here because the fpc should be computed when the sample is selected. If the sampling fraction $n/N$ is substantial, for example, bigger than 0.05, this will noticeably reduce variances. So, it is important to take advantage of this by specifying a field to hold the fpc. The fpc field can be specified in Stata in any of three ways:

1. with a value between 0 and 1, in which case the field is interpreted as the sampling rate $n/N$;

2. with the number of units sampled, $n$; or

3. with an integer greater than $n$, in which case the field is interpreted as containing $N$.

The next three examples show how to select various types of simple random samples. The code for all three examples is in the file `ex.2.1_2.2_2.3_srs.do`.

**Example 2.1: Select an srswor of fixed size.** In this example, we select an srswor of size 100 from `nhis.dta`, a dataset from the U.S. National Health Interview Survey. The base weight for each case is $N/n = 3911/100 = 39.11$ and is stored in the field `srswt`. The fpc is stored in `fpc` as the value of $N$. The `count` option of the `sample` command means that 100 is interpreted as the number of cases to be selected—not as a percentage of the population to sample.

```
. use http://www.stata-press.com/data/svywt/nhis
. generate srswt = _N/100
. generate fpc = _N
. sample 100, count
```

■

**Example 2.2: Select an srswor at a fixed rate.** The `sample` command will also select an srswor at a specified rate $f$, for example, a 10% sample, if the `count` option is omitted. In this case, the weight could be computed as $1/f$ or as $N/n^*$, where $n^*$ is the number of cases actually selected if $fN$ is not an integer. This example selects a 10% srswor and computes the weight as $1/0.10 = 10$. The fpc is stored in `fpc`.

```
. use http://www.stata-press.com/data/svywt/nhis, clear
. generate srswrwt = 1/0.10
. generate fpc = 0.10
. sample 10
```

∎

If there is nonresponse, then $n_{\text{ER}}$ units will respond rather than the $n$ initially selected. In that case, the fpc used to estimate the SE should be treated as $1 - n_{\text{ER}}/N$ because data are obtained only for the respondents. This does require assuming that response is just another stage of equal probability sampling—an issue that we discuss further in chapter 3.

**Example 2.3: srswr example.** If you want an srswr, use the `bsample` command. For example, the following code will make 100 random draws with replacement from `nhis.dta`. The pwr weight is computed as $N/100$. You can also use this method to select standard bootstrap samples. A unit will occur in the resulting dataset as many times as it is selected—1, 2, 3, etc. Note that an fpc is not appropriate for a with-replacement sample because this method is equivalent to sampling from an infinite population.

```
. use http://www.stata-press.com/data/svywt/nhis, clear
. generate srswrwt = _N/100
. bsample 100
```

∎

If you want to be able to reproduce the same sample (regardless of whether it is with- or without-replacement), a random-number seed should be set before sampling with a command like the following:

```
. set seed 339487731
```

The seed can be any number between 0 and $2^{31} - 1$. Setting a seed value is good practice to ensure that the same sample is selected if the program is resubmitted to include, for example, additional quality checks.

A related case is equal-probability, systematic sampling. In that case, the frame is sorted in some order, a random starting place is selected, and that unit plus every $(N/n)$th unit thereafter is selected. The value of $N/n$ is called a "skip interval". The properties of this method depend on the sort order. If the order is random, then the method is equivalent to srswor. If the list is sorted by some characteristic related to study variables, for example, schools sorted by enrollment, establishments sorted by number of employees, or persons sorted by gender, then the sample is said to be "implicitly stratified". (In contrast, "explicit stratification" occurs when samples are selected within independent groups as discussed next.) Systematic sampling spreads the sample across the range of the sorting characteristics. Regardless of the sort order, the selection probability is $n/N$ and the base weight is $N/n$.

**Example 2.4: Systematic sample.** The example below (in `ex.2.4_sys.do`) selects a systematic sample of $n = 100$ persons from `nhis.dta` after sorting by categorical age (`age_r`) and whether a person is Hispanic (`hisp`). The skip interval is $N/n = 3911/100 = 39.11$. The command `generate low = int(runiform()*skip)+1` generates a random integer starting point between 1 and $N/n + 1$.[2] Records with sequential counters in the ordered file less than `low` are dropped. The command `generate y = mod(newsno, 39)` is 0 for those remaining records whose counters (minus 1) are multiples of 39. Those records are then retained with `drop if y != 0`. The base weight `bwt` is computed as the full population size `Npop` divided by the number actually selected, `n_select`.

```
. use http://www.stata-press.com/data/svywt/nhis, clear
. set seed 37567298
. generate Npop = _N
. generate n = 100
. generate skip = Npop/n
. generate low = int(runiform()*skip)+1
. sort age_r hisp
. drop if _n < low
. generate newsno = _n - 1
. generate y = mod(newsno, 39)
. drop if y != 0
. generate n_select = _N
. generate bwt = Npop/n_select
```

You can verify that the sample is distributed about the same as the population on `age_r` $\times$ `hisp`. There are many other ways of selecting systematic samples other than the one above. Särndal, Swensson, and Wretman (1992) give some options.

◼

**Stratified simple random sampling**. Design strata can be created from one or more characteristics of interest for the study for administrative reasons, analytic objectives (for example, minimum sample size to meet precision requirements for a certain domain), or simply to avoid samples that are poorly distributed across groups in the population (as can occur with simple random samples). For example, a study of employee satisfaction within a particular company could include a single-stage sample of employees stratified by and selected independently within business unit.

If a simple random sample of $n_h$ units is selected without replacement within stratum $h$ $(h = 1, \ldots, H \geq 2)$ of population size $N_h$ with sampling fraction (or rate) $n_h/N_h$, the base weight would be $d_{0i} = N_h/n_h$. If the sample within each stratum were srswr, then the pwr base weight would also be $N_h/n_h$.

---

2. The `int()` function gives the integer lower bound so that `int(runiform()*skip)+1` is between 1 and 1 plus the greatest integer less than the skip interval, that is, between 1 and 40 in this example.

Note that the sampling rates are allowed to and typically do differ across strata. For example, certain strata may have relatively higher or lower sampling rates to ensure the required sample size meets the analytic goals for the study. Sampling rates may also differ when population variances differ across the strata as in the case of Neyman allocation (see, for example, Valliant, Dever, and Kreuter [2013]).

The `sample` command will select stratified simple random samples without replacement (stsrswor), but the sample size or the sampling rate must be the same in every stratum. For example, this command will select an stsrswor of five persons per stratum from the `nhis.dta` file using four strata defined by the variable `educ_r`: $1 =$ high school, general education development (GED) degree, or less; $2 =$ some college; $3 =$ bachelor's or associate's degree; and $4 =$ master's degree and higher:

```
. sample 5, count by(educ_r)
```

The code above both selects the sample and extracts the 20 selected observations from `nhis.dta`. Requiring that the sample sizes or rates be the same in each stratum is limiting. Writing a special-purpose do-file is more flexible.

**Example 2.5: Stratified simple random sampling with varying stratum sample sizes.** The code (in `ex.2.5_stsrs.do`) below will select an stsrswor from the population in `nhis.dta`. The stratum population counts are stored in Nh, which is a $4 \times 1$ matrix. The sample sizes are in nh, a $1 \times 4$ matrix equal to $\mathbf{n}_h = (3, 4, 5, 6)$. A `foreach` loop is used to select the stratum samples. After the samples are selected, the base weights are computed as Nh/nh and stored as `stsrswt`.

```
. set seed 339487731
.       * sort population
. sort educ_r

. levelsof educ_r, local(edlev)
1 2 3 4

. tabulate educ_r, matcell(Nh)
. matrix nh = (3, 4, 5, 6)
. foreach i of local edlev {
  2.        local n = nh[1,`i´]
  3.        sample `n´ if educ_r == `i´, count
  4. }
.       * sort sample
. sort educ_r
. generate stsrswt = Nh[1,1]/nh[1,1] if educ_r == 1
. replace stsrswt = Nh[2,1]/nh[1,2] if educ_r == 2
. replace stsrswt = Nh[3,1]/nh[1,3] if educ_r == 3
. replace stsrswt = Nh[4,1]/nh[1,4] if educ_r == 4
```

∎

Another option for selecting stratified and many other types of samples is the `sampling` package in R (Tillé and Matei 2016). R routines can be called from Stata using `rsource` written by Newson (2014). The website for this book at http://www.stata-press.com/books/survey-weights has several downloadable examples that use `rsource`.

**Probability proportional to size sampling**. In probability proportional to size (pps) sampling, units are assigned a measure of size (mos). The sample is then selected in such a way that the selection probability of a unit is proportional to its mos. This type of sampling can be used in single-stage samples, for example, of business establishments, or in multistage samples at the first or later stages. Size measures for pps sampling are ideally linearly related to key information collected from the sample. These values can be, for example, 1) a function of a unit's size, such as $\sqrt{x}$, identified as the optimal form of the size measure to minimize variance (see, for example, Valliant, Dorfman, and Royall [2000]); 2) a simple measure of the cluster size, for example, the number of students in a school; or 3) a function of group counts within a cluster, which are used to create a composite size measure (Folsom, Potter, and Williams 1987). Composite size measures have been used, for example, in education surveys where the analytic objectives require a minimum number of participating students by race. This type of size measure will effectively "up weight" schools with a higher proportion of students with rarer characteristics.

**Example 2.6: pps sample of fixed size.** The example below selects a pps sample of size 30 from the Survey of Mental Health Organizations (SMHO) population in the `PracTools` R package (Valliant, Dever, and Kreuter 2016). The code is in `ex.2.6_pps.do`. The mos is computed by adding 5 to the number of inpatient beds in each hospital and then taking the square root. The 5 is added because some outpatient hospitals have 0 beds and would have no chance of selection without recoding beds to a nonzero value. The selection probabilities are computed as $nx_i/t_{Ux}$, where $x$ is the recoded mos, $t_{Ux}$ is the sum of $x$ over the whole population, and $n$ is the sample size. The `samplepps` command (Jenkins 2005) selects a pps sample by randomly ordering the population and then selecting a systematic sample with probability proportional to the recoded mos, nBEDS.

```
. use http://www.stata-press.com/data/svywt/smho.n874.dta, clear
. set seed 339487731
. generate nBEDS = BEDS
. replace nBEDS = BEDS + 5 if BEDS < 5
. replace nBEDS = sqrt(nBEDS)
. generate tBEDS = 6383.244
. generate sel_prob = 30*nBEDS / tBEDS

. summarize sel_prob
```

| Variable | Obs | Mean | Std. Dev. | Min | Max |
|---|---|---|---|---|---|
| sel_prob | 874 | .034345 | .025342 | .0105091 | .173129 |

```
. samplepps sel_sw, ncases(30) size(nBEDS)
. keep if sel_sw == 1
. generate bwt = 1/sel_prob
```

The final line of the code above computes the base weights as the inverse of the selection probability for each sample unit.

Two other commands that select pps samples using specialized algorithms are `ppschromy` and `ppssampford` (Mendelson 2014a,b).

∎

Selection probabilities are typically greater than zero and less than one. For pps sampling, however, these values can sometimes exceed one. Units with such values are called "certainties", "self-representing", or "take-alls". A few options available to sampling statisticians for this instance are 1) in a single-stage sample, set the probability (and base weight) of the certainty case to one and select the remaining sample from the frame after recalculating the probabilities with the certainty removed; in multistage samples, either 2) reclassify the primary sampling unit (PSU) as a first-stage stratum and select from a new set of PSUs created from the elements of the original cluster, or 3) allow the PSU to be selected multiple times with independent second-stage samples selected for each. Choosing a specific approach to handle certainty first-stage units will dictate the form of the base weights.

All of this is to say that creation of the size measure and the selection probabilities are important to ensure the accuracy of the base weights. We discuss quality assurance and control procedures in depth in chapter 8.

**Multistage sampling**. Suppose that first-stage cluster $i$ (also known as a PSU) is selected within probability $\pi_i$ and that an element $j$ within cluster $i$ is selected with conditional probability $\pi_{j|i}$. We call the second-stage probability "conditional" because it is conditioned on cluster $i$ being selected. The (unconditional) base weight would then be $d_{0i} = \left(\pi_i \pi_{j|i}\right)^{-1}$. For example, suppose that a sample of students is selected by randomly sampling schools at the first stage and students at the second. Let $E_i$ represent the number of students in school $i$. A total of $m$ schools is selected for the study, each with probability proportional to the size of the student body: $\pi_i = mE_i/E_{U+}$ where $E_{U+}$ is the total number of students across all schools in the frame. If the same number, $\bar{n}$, of students is selected via equal-probability sampling in each school, then the selection probability of a student is $mE_i/E_{U+} \times \bar{n}/E_i = m\bar{n}/E_{U+}$. The weight for each student is the same, $E_{U+}/(m\bar{n})$.

The two-stage sample above is an example of what is known as a "self-weighting" sample, that is, each sample element has the same weight. A simple random sample is also self-weighting, because each weight is $N/n$. A self-weighting sample can be selected in any number of stages. Although this may simplify point estimation of means, totals, and other quantities, other features of the sample design such as stratification and clustering still must be considered when estimating variances. In addition, the equal-weights feature may be lost if one makes nonresponse or calibration adjustments, as discussed in chapters 3 and 4, respectively.

A multistage sample can be selected sequentially, which is the standard approach needed for all software to date. First, the sample of PSUs is selected. Then, the selected PSUs can be treated as strata from which the second-stage units are selected. Selecting subsequent stages, if any, would proceed in the same way.

**Example 2.7: Two-stage sampling.** In the following, we take a sample of 10 clusters and 5 persons per cluster. In the first stage, the `TRACT` variable defines the clusters. We use the `contract` command to generate a dataset consisting of the unique levels of `TRACT` and a new variable named `Ni`, containing the count of persons within each tract from the original data. We then generate `pi1` to hold the selection probabilities of the clusters and use the command `samplepps` to select the pps sample of 10 tracts, keep only the variables and observations for the selected clusters, and save the contracted dataset as `mdarea-sam-psu-pps.dta`.

```
. set seed 339487731
. use http://www.stata-press.com/data/svywt/mdarea, clear
. contract TRACT, freq(Ni)
. summarize Ni, meanonly
. generate double pi1 = 10*Ni/r(sum)
. label var pi1 "Stage 1 sampling probabilities"
. samplepps sel, ncases(10) size(pi1)
. keep if sel == 1
. keep TRACT pi1


. describe
Contains data from http://www.stata-press.com/data/svywt/mdarea.dta
  obs:            10                          Written by R.
  vars:            2
  size:          160

                storage   display    value
variable name    type     format     label    variable label

TRACT           double    %9.0g                TRACT
pi1             double    %10.0g               Stage 1 sampling probabilities

Sorted by: TRACT
     Note: Dataset has changed since last saved.

. save mdarea-sam-psu-pps
```

Using the contracted dataset, we can select the clusters from the full dataset.

```
. use http://www.stata-press.com/data/svywt/mdarea, clear

. merge m:1 TRACT using mdarea-sam-psu-pps
    Result                         # of obs.

    not matched                      349,739
        from master                  349,739   (_merge==1)
        from using                         0   (_merge==2)
    matched                           54,258   (_merge==3)


. keep if _merge == 3
. drop _merge
```

In the second stage, we sample five persons from each tract. We begin by sorting the data on TRACT and storing the conditional sampling probabilities for the second stage in a new variable named pi2. Similarly to example 2.5, we then loop over the tract levels and use sample to select five persons from within each tract.

```
. sort TRACT
. by TRACT : generate double pi2 = 5/_N
. label var pi2 "Stage 2 sampling probabilities"
. levelsof TRACT, local(TRACTlev)
702100 702402 702602 707001 730700 731103 731310 731311 740301 750801
. foreach i of local TRACTlev {
  2.         quietly sample 5 if TRACT == `i´, count
  3. }
. generate double bwt = 1/(pi1*pi2)
. summarize bwt
    Variable |       Obs        Mean    Std. Dev.       Min        Max

         bwt |        50     8079.94     9.19e-13    8079.94    8079.94
```

We finish up by putting the base weights in variable bwt, computed as $(\pi_{i1}\pi_{j|i})^{-1}$. We use summarize on bwt to verify that the sample is self-weighting because each person has a weight of 8079.94.

■

**Software**. Many software packages exist to select single-stage random samples from a frame, including the Stata sample and samplepps commands, the R sampling and survey packages, and the SAS Proc SurveySelect. This is especially helpful when using more complex methods than discussed above such as balanced sampling (Deville and Tillé 2004) and other iterative methods. You may still need to "grow your own software" to address your particular needs such as sampling based on geographic information system (GIS) data (Eckman, Himelein, and Dever Forthcoming).

To date, multistage random samples must be selected one stage at a time for most software under our purview. Some, but not all, will output the base weights or the selection probabilities used to calculate the base weights for each stage of the design. Otherwise, it is wise to calculate the selection probabilities as soon as the sample is drawn. One should verify the selection probabilities and the base weights before releasing the sample to the data collectors (see chapter 8).

### Release sample in replicates

Sample designs are constructed with estimated or assumed response rates. If the confidence in these rates is lacking, researchers may select more samples than they expect to use. Then, only the portion of the sample required to meet set targets for the study (for example, minimum number of completed interviews) is released for data collection (that is, fielded) to contain cost. Consequently, it is necessary for random sampling to mirror the original design.

The enhanced sample is randomly divided into release replicates so that each replicate is a miniature version of the full sample. For example, if an stsrswor is selected, the sample units in each stratum are randomly assigned to a replicate to maintain the relative distribution across the strata. The overall size of the replicates need not be the same; only random sampling is critical.

Once constructed, the researcher will field a subset of the replicates and monitor progress closely. If the projected targets are insufficient, then additional replicates are released for recruitment. Otherwise, a certain number of replicates are never fielded.

Because the replicates were created by randomly sampling from the original set of units, those units actually fielded are a "legitimate" subsample. The statistician simply adjusts the original base weights to account for the subsampling. Continuing with our stsrswor example above, the adjusted base weight within stratum $h$ ($h = 1, \ldots, H \geq 2$) is calculated as

$$
\begin{aligned}
d_{0i}^* &= a_{0i}^* d_{0i} \\
&= \frac{N_h}{a_{0h}}
\end{aligned}
$$

where $a_{0i}^{*-1} = a_{0h}/n_h$, the subsampling rate associated with the released portion of the full sample in stratum $h$, $a_{0h}$ is the number of sample units released, and $d_{0i} = N_h/n_h$, the original stratum-specific base weight. The base weight for sample units never released is set to zero, that is, $a_{0i}^* = 0$. The weight $d_{0i}^*$ is the new base weight used in subsequent adjustments such as for unknown eligibility, which we discuss next.

## 2.2   Adjustments for unknown eligibility

In some surveys, it may not be possible to determine whether some sample units are members of the target population and thus eligible for the study. We gave a few examples in section 1.7. Another is a telephone survey of households in which some numbers are never answered, and no voicemail message is heard to indicate that the number belongs to a household or is associated with an activated cellular telephone number. A second example is a survey of the membership of a professional organization in which an out-of-date frame is used. Suppose that some persons cannot be contacted because they have left the organization or have moved, and no current address can be obtained. The eligibility of those persons is unknown, unless we are willing to make an arbitrary decision that all are ineligible (IN), all are eligible nonrespondents (ENR), or we are confident in our ability to impute their eligibility status based on frame and data collection information. If they are classified as unknown (UNK), then one approach is to make a simple weight adjustment to the units with known eligibility. This is step 2 in figure 1.2.

The adjustments for unknown eligibility are usually fairly simple because very little may be known about the cases whose eligibility cannot be determined. The simplest would be to form groups of sample units (classes or cells) based on information known for all units whether or not they are eligible. The steps to do this are the following:

1. Form $b = 1, \ldots, B$ classes based on frame information known for all cases. Classes may cut across design strata. In practice, eligibility adjustment and nonresponse adjustment classes may be the same.

2. Let $s_b$ be the set of sample units in class $b$, regardless of eligibility or response status, and let $s_{b,\text{KN}} = s_b \cap s_{\text{KN}}$ be the set with known eligibility in class $b$. The symbol $\cap$ identifies the set of units in $s_b$ and in $s_{\text{KN}}$ (that is, the intersection).

3. The unknown eligibility adjustment for sample units in class $b$ is
   $a_{1b} = (\sum_{i \subset s_b} d_{0i})/(\sum_{i \in s_{b,\text{KN}}} d_{0i})$, where $d_{0i}$ is base weight.

4. The adjusted weight for unit $i$ in $s_{b,\text{KN}}$ is $d_{1i} = a_{1b} d_{0i}$. The factor $1/a_{1b}$ functions as an estimate of the probability of having known status (eligible or ineligible). The weights for the remaining units in class $b$, those with unknown eligibility, are set to 0; that is, $a_{1b} = 0$ for $s_{b,\text{UNK}} = s_b \cap s_{\text{UNK}}$.

A typical rule of thumb is to ensure that each weighting class contains at least 50 sample units. Another criterion would be to require the proportion of unknown cases to be less than some cutoff; however, there appears to be no commonly agreed upon standard for this. We revisit these topics in detail in chapter 3 for nonresponse weighting classes.

# 3   Adjustments for nonresponse

Nonresponse (NR) occurs in most surveys regardless of the units being surveyed—persons, businesses, schools, hospitals, etc. The general idea in nonresponse adjustment is to increase the weights of the respondents to make up for the fact that some units did not provide data. The method of weight adjustment covered in this chapter is called weighting class adjustment. Several techniques, covered in subsequent sections, can be used for forming classes (also referred to as cells). To understand the methods that practitioners use, we need a formal way of thinking about how nonresponse occurs. Response to a survey can be thought of as either a deterministic or stochastic process. These can be summarized as follows:

1. **Deterministic**—each eligible unit in the population will have a choice to respond if asked to participate. The choice is not random, so units could be sorted a priori into respondents and nonrespondents.

2. **Stochastic**—each unit has some nonzero probability of responding. When asked to participate, a unit makes a random choice to cooperate or not.

Suppose that $\widehat{\bar{y}}_R$ is the estimate, based on sample respondents, of a population mean but that it actually estimates the population mean of respondents, $\overline{Y}_{\mathrm{ER}}$—not the mean for all units in the population. The bias of the estimated mean when there is deterministic response is

$$\mathrm{NR.bias}_D\left(\widehat{\bar{y}}_R\right) = M\left(\overline{Y}_{\mathrm{ER}} - \overline{Y}_{\mathrm{ENR}}\right)/N \qquad (3.1)$$

where $\overline{Y}_{\mathrm{ENR}}$ is the true mean for the nonrespondent population, and $M/N$ is the population nonresponse rate calculated as the ratio of the nonrespondent population size, $M$, over the population size, $N$. In the deterministic formulation, there is a bias if the population mean for the respondents is different from that of the nonrespondents. The idea behind a weighting class adjustment method is to try and group units together in such a way that the class means for respondents and nonrespondents are equal, that is, $\overline{Y}_{\mathrm{ER}} = \overline{Y}_{\mathrm{ENR}}$. Note that the bias in (3.1) is also zero if there is 100% participation, that is, $M = 0$; however, this occurrence is nonexistent nowadays.

The deterministic approach seems simplistic because units probably cannot be cleanly separated in groups, much like design strata, that will always or will never respond. The stochastic approach underlies most of the nonresponse adjustment techniques used in practice. As shown below, the algebra is more elaborate in the stochastic approach.

Define two indicators for being in the sample and responding:

$$I_i = \begin{cases} 1 & \text{if unit } i \text{ selected for sample} \\ 0 & \text{if not} \end{cases}$$

$$R_i = \begin{cases} 1 & \text{if unit } i \text{ responds given that it is in the sample} \\ 0 & \text{if unit } i \text{ does not respond} \end{cases} \qquad (3.2)$$

The probability of being in the sample is $\Pr(I_i = 1) = \pi_i$, while the probability of responding given that unit $i$ is in the sample is $\Pr(R_i = 1 \mid I_i = 1) = \phi_i$. Rosenbaum and Rubin (1983) call $\phi_i$ the propensity score for unit $i$. There could be bias if $\phi_i = 0$ for some units; that is, some units are "hard-core" nonrespondents who would never participate in a survey. If all units have some nonzero probability of responding, then it may be possible to produce estimates that are, in some statistical sense, unbiased.

Suppose $d_{0i} = 1/\pi_i$ is the base weight we assign to unit $i$ and consider the survey-weighted estimator of a mean: $\widehat{\overline{y}}_R = \sum_{i \in s_{\text{ER}}} d_{0i} y_i / \sum_{i \in s_{\text{ER}}} d_{0i}$. Under the "quasirandomization" setup, where sampling and responding are both considered to be random, Kalton and Maligalig (1991) showed the NR bias of $\widehat{\overline{y}}_R$ is

$$\text{NR.bias}_S\left(\widehat{\overline{y}}_R\right) \doteq \frac{1}{N\overline{\phi}} \sum \left(y_i - \overline{Y}_U\right)\left(\phi_i - \overline{\phi}\right) \qquad (3.3)$$

where $\overline{\phi}$ is the average population probability of responding and $\overline{Y}_U$ is the true mean of $y$ in the universe. Consequently, the bias depends on the covariance of the $y$ variable and its response propensity. If $y_i$ and $\phi_i$ are unrelated, there is no bias and nonresponse does not need to be corrected, at least when estimating a mean.

However, we usually fear that the chances of responding may be related to the $y$'s and that we need to do something to reduce or eliminate bias. One property that we could hope to achieve is unbiasedness when we average over both the random selection of the sample and the random way in which units choose to respond or not respond. Suppose $w_i^*$ is the weight we assign to unit $i$ after nonresponse adjustment and consider this estimator of a total calculated from the set of respondents, $s_{\text{ER}}$:

$$\widehat{t}_y = \sum_{i \in s_{\text{ER}}} w_i^* y_i$$

The theoretical average (expected value) of this estimator over the sampling and response mechanisms is

$$\begin{aligned} E_I E_R\left(\widehat{t}_y\right) &= E_I E_R\left(\sum_{i \in U} R_i I_i w_i^* y_i\right) \\ &= \sum_{i \in U} w_i^* y_i \left\{E_I E_R\left(R_i I_i\right)\right\} \end{aligned}$$

where $E_I$ is the expectation (average) with respect to the sampling design (as discussed in section 1.3), and $E_R$ is the expectation taken with respect to the response mechanism. If we can make $w_i^* = 1/E_I E_R\left(I_i R_i\right)$, this reduces to the population total, $\sum_{i \in U} y_i$, the desired result. Because $E_R E_I\left(I_i R_i\right) = E_I\left\{I_i E_R\left(R_i \mid I_i\right)\right\} = \pi_i \phi_i$, where $E_R\left(R_i \mid I_i\right)$ is $E_R$ conditioned on the chosen sample, the weight would be $w_i^* = \left(\pi_i \phi_i\right)^{-1}$. This, of course, requires that both $\pi_i$ and $\phi_i$ be nonzero.

# Patterns of missing data

Three other ideas are useful when thinking of nonresponse or how data can be missing generally: missing completely at random (MCAR), missing at random (MAR), and not missing at random (NMAR) (also known as nonignorable nonresponse). These are due to Little and Rubin (2002); we sketch their meanings below. The definition of each term requires us to think of a third distribution—a model for an analysis variable $Y$. In fact, if $K$ analysis variables are collected on each unit, $\mathbf{y}_i = (y_{i1}, y_{i2}, \ldots, y_{iK})$ must be considered. Also, suppose that there is a set of auxiliary variables $\mathbf{x}_i = (x_{i1}, x_{i2}, \ldots, x_{ip})$ available for each sample unit whether or not it responds. These can be items like age, race, and sex in a household survey or type of business and number of employees in a business establishment survey. The $x$'s can also include information used in the sample design, like region of the country and type of area (urban, suburban, or rural) or observations reported by interviewers about the condition of a neighborhood.

These observational data are referred to as paradata and are discussed in Kreuter, Couper, and Lyberg (2010). However, some caution is required when using some kinds of paradata for nonresponse adjustment. Kreuter and Olson (2011) illustrate that if, say, trash on the streets of a neighborhood or difficulty in finding someone at home are unrelated to the analysis variables collected in a survey, then using those paradata in nonresponse adjustment may do more harm than good. Using irrelevant data may just inject pointless variability into estimates without correcting any bias.

**Missing completely at random (MCAR).** If the probability of response $\phi_i$ does not depend on $\mathbf{y}_i$ or $\mathbf{x}_i$, then the missing data are MCAR. If everyone has the same probability of responding, $\phi$, then any set of nonrespondents are MCAR. In other words, the respondent set is a random subsample of the population, and nonresponse adjustments are unnecessary. Researchers who use weights to calculate estimates seldom assume MCAR.

**Missing at random (MAR).** If the probability of response does not depend on $\mathbf{y}_i$ but does depend on some of or all the auxiliaries $\mathbf{x}_i$, then the missing data are MAR. In this case, a model for response can be formed that depends on $\mathbf{x}_i$ because we know the auxiliaries for both respondents and nonrespondents. MAR, as Lohr (2010) notes, is sometimes called "ignorable nonresponse", meaning that if the response mechanism is modeled correctly and adjustments for nonresponse are made, then inferences to the population are possible.

**Not missing at random (NMAR).** If the chances of responding depend on one or more analysis variables (that is, the $\mathbf{y}$'s) and this dependence cannot be eliminated by modeling response based on $\mathbf{x}$'s that are known for both respondents and nonrespondents, then the data are NMAR. Suppose, in a personnel survey, we were able to model response as a function of business unit of a company, pay grade, etc. plus an analysis variable that rates whether employees thought there was a clear link between performance rating and pay. If the coefficient on the rating variable was significant, this would be evidence of NMAR. The practical problem with fitting this kind of model is that the rating for the nonrespondents will not be available. Consequently, NMAR is difficult or impossible to detect in a cross-sectional survey and is the worst situation to have when trying to adjust for nonresponse bias. (An exception is a panel survey where values of analytic variables may be available from previous time periods for current period nonrespondents.)

## 3.1   Weighting class adjustments

Judging from expression (3.3), the nonresponse bias can be eliminated if responding and nonresponding units can be grouped into classes where either 1) all units have the same probability of responding or 2) all have the same mean of $y$ within each group. Because the $y$ values are unknown for nonrespondents, achieving goal 1 is typically more feasible. For the methods described in this section and the subsequent ones, the values of the auxiliary variables used to form classes must be known for both the respondents and the nonrespondents. In some surveys, this may severely restrict the options for NR adjustment.

In this section, we cover the mechanics of using classes to make nonresponse adjustments. There are different ways of forming classes. The simplest is to identify some variables (like age, sex, and other race in a household panel survey) that are known to be related to the likelihood of responding, the mean of some important $y$'s, or both. A critical requirement is that the values of classification variables be known for both respondents and nonrespondents. These variables are then categorized and crossed to form classes. More formal ways of forming classes are to use estimated propensity scores, described in section 3.2 or a classification algorithm, as discussed in section 3.3.

We index the classes by $c = 1, \ldots, C$. If all units in a class have the same covariate values, $\mathbf{x}_c$, and response propensity for a unit is a function of $x_c$, then $\phi_i = \phi(\mathbf{x}_c)$ for all units in $c$. Denote the set of eligible sample units in class $c$ as $s_{c,E}$ of size $n_{c,E}$, the set of $n_{c,\text{ER}}$ eligible respondents as $s_{c,\text{ER}}$, and the set of eligible nonrespondents by $s_{c,\text{ENR}}$ with size $n_{c,\text{ENR}} = n_c - n_{c,\text{ER}}$. One type of nonresponse adjustment for units in class $c$ is computed using the unknown-eligibility adjusted weights given in section 2.2:

$$a_{2c} = \frac{\sum_{i \in s_{c,E}} d_{1i}}{\sum_{i \in s_{c,\text{ER}}} d_{1i}} \tag{3.4}$$

This is the ratio of the sum of the input weights for all eligible units in the class to the sum of the input weights for the eligible respondents in that class. Because the sum of weights can be interpreted as an estimate of the total number of units in the population in whatever class is included in the summation, $a_{2c}$ is the ratio of the estimated total number of eligible units in class $c$ to the estimated number of eligible respondents in that class. Thus, the inverse of $a_{2c}$ is an estimate of the response rate in the class if a census were conducted. If the grouping has succeeded in putting together units that have a common response probability, then $1/a_{2c}$ is an estimate of that probability.

Also, this would imply that after grouping into the classes, the data are MAR because the response probability would depend on auxiliaries (the class indicators) but not on the $y$'s.

The resulting adjustment $a_{2c}$ is applied only to the eligible respondents in class $c$. The adjustment is set to 0 for the unknowns or known eligible nonrespondents, $s_{\text{UNK}} \cup s_{\text{ENR}}$, and to 1 for the units known to be ineligible, $s_{\text{IN}}$. The weight for unit $i$ in the initial sample, after the adjustments for unknown eligibility and nonresponse, is then

$$d_{2i} = \begin{cases} d_{1i}a_{2c} & i \in s_{c,\text{ER}} \\ d_{1i} & i \in s_{\text{IN}} \\ 0 & i \in s_{\text{UNK}} \cup s_{\text{ENR}} \end{cases}$$

$$= \begin{cases} d_{0i}a_{1b}a_{2c} & i \in s_{b,KN} \cap s_{c,\text{ER}} \\ d_{0i}a_{1b} & i \in s_{b,KN} \cap s_{\text{IN}} \\ 0 & i \in s_{\text{UNK}} \cup s_{\text{ENR}} \end{cases}$$

The eligible respondents get both the adjustment for unknown eligibility and the nonresponse adjustment. Known ineligibles ($s_{\text{KN}} \cap s_{\text{IN}}$) get only the unknown eligibility adjustment. Unknowns ($s_{\text{UNK}}$) and eligible nonrespondents ($s_{\text{ENR}}$) drop out.

The $a_{2c}$ adjustment does not necessarily have to use the $d_{1i}$ weights. As pointed out by Little and Vartivarian (2003), if all units in a nonresponse adjustment class have the same response probability, then an unweighted adjustment, $a_{2c} = n_{c,E}/n_{c,\text{ER}}$, will be unbiased with respect to the response model and can give more stable NR adjustments, that is, values of $a_{2c}$ that do not vary as much as they would if the $d_{1i}$ weights were used.

**Example 3.1: Simple nonresponse class adjustments.** The following is a tabulation of the (weighted) proportions of persons responding to an income question in `nhis.dta` from the U.S. National Health Interview Survey (see `ex.3.1_agenr.do`). The dataset is also available in the R package `PracTools` (Valliant, Dever, and Kreuter 2016). (For illustrative purposes, we use the item (income) response indicator and respondent data in place of a unit (survey) response indicator and full sample.) The variable `resp`, coded 1 or 0, specifies whether a person is a respondent. The proportions responding are shown for six different age groups. Noting the differences in the unweighted response rates across age groups, the variable `age_r` might be a candidate for forming NR adjustment cells. For example, the NR weight adjustment for ages 18–24 could be 1/0.6831, the inverse of the proportion responding. However, response may depend on other factors, in which case this adjustment would be too simplistic.

```
. use http://www.stata-press.com/data/svywt/nhis

. label define age_r_lab 3 "18-24" 4 "25-44" 5 "45-64" 6 "65-69" 7 "70-74"
> 8 "75+"

. label values age_r age_r_lab
```

```
. svyset psu [pweight=svywt], strata(stratum)

. svy: tabulate age_r resp, row
(running tabulate on estimation sample)
Number of strata   =         87                  Number of obs      =      3,911
Number of PSUs     =        174                  Population size    = 12,386,519
                                                 Design df          =         87

----------------------------------
          |          resp
    age_r |       0       1  Total
----------+-----------------------
    18-24 |   .3169   .6831      1
    25-44 |   .2496   .7504      1
    45-64 |   .3001   .6999      1
    65-69 |   .3881   .6119      1
    70-74 |   .3085   .6915      1
      75+ |    .399    .601      1
          |
    Total |   .2954   .7046      1
----------------------------------
  Key:  row proportion
  Pearson:
    Uncorrected   chi2(5)         =   40.0371
    Design-based  F(4.59, 399.14) =    6.8861     P = 0.0000
```

■

If response rates differed by, say, age, race, and sex, using cells defined by the three-way cross of those variables would be reasonable. However, in some applications, some of the cells may have few, if any, eligible sample units or sample respondents. The inverse of the ratio in (3.4) is an estimated proportion and can be unstable, especially if $n_{c,E}$ is small. Generally, practitioners use rules of thumb rather than formal calculations to set minimum sample sizes of eligible units per cell. Requiring $n_{c,E}$ to be at least 30 is common. Any cell with fewer than some minimum number of units would be combined with another when making an NR adjustment.

Given the mechanics, the question then becomes: are there more efficient ways of forming classes? We cover two methods in the next two sections.

## 3.2 Propensity score adjustments

As noted at the beginning of this chapter, we will obtain an estimator of a total that is unbiased over the combined sampling or response process if the weight is $d_{2i} = 1/(\pi_i \phi_i)$. If $\phi_i = \phi(\mathbf{x}_i)$, we can try to model the response probabilities if we measure the covariates on all eligible sample units.

Positing that the chance of responding depends only on covariates that we have available is a strong assumption that may or may not be true. There are problems when units are not MAR or MCAR. For example, if $\phi_i = \phi(y_i)$, we do not have $y$'s for nonrespondents ($R = 0$). If the nonrespondents follow a different model from the respondents, we will not know it. Another possible problem is some covariates that predict response may be unobserved or unused. Then, $\phi_i = \phi(U_i)$ where $U_i$ contains the unmeasured covariates or measured covariates incorrectly omitted from the model. For example, suppose that response in a school survey depends on grade range, school size, and a measure of socioeconomic status of the students, but we omit socioeconomic status. A common situation would be that response depends on a covariate that is not measured on either the respondents or the nonrespondents.

It is probably safe to assume that we are operating with inadequate information, but, in practice, model parameters must be estimated based on what is known for both respondents and nonrespondents. You do the best you can based on the available information. One approach is to fit a binary regression model for the response indicators $R_i$ in (3.2). The expected value of the indicator is

$$E_R\left(R_i|\,I_i=1\right) = \Pr\left(R_i=1|\,I_i=1\right) = \phi\left(\mathbf{x}_i\right) \tag{3.5}$$

This is the conditional probability of response given that a unit is selected for the sample that is a function of the auxiliary set $\mathbf{x}_i$. This also has a bearing on whether to use base weights in fitting the model, as discussed below.

The probability of response can be estimated using a binary regression model. Given the estimated probabilities or propensities, two options for adjustment are

1. Use the individual estimated propensities, $\widehat{\phi}\left(\mathbf{x}_i\right)$. If the response model is correctly specified, this option provides unbiased estimates with respect to the response mechanism. To avoid extreme adjustments, the estimates can be bounded.

2. Sort the propensities and group them into classes. Unlike the weighting class method in section 3.1, the propensity score approach easily allows the use of multiple categorical and even continuous variables to address nonresponse. When classes are used, a single weight adjustment is calculated within each class to reduce the variability that could result from using individual estimated propensities for adjustment (see, for example, Haziza and Lesage [2016]).

Three well-known binary regression models are logistic, probit, and complementary log-log. In many applications, these models fit equally well for a given set of covariates, so the choice of which model to use is not a challenge.

**Example 3.2: Propensity classes.** This example uses logistic regression to estimate the propensities of responding to an income question in `nhis`, the dataset from the U.S. National Health Interview Survey. A snip of the code and output is listed below. The do file for this example is `ex.3.2_logisticnr.do`.

The variable `resp`, coded 1 or 0, specifies whether a person is a respondent. The probability of responding to the income question is modeled based on the person's age, an indicator for whether a person is Hispanic (`i.hisp`), an indicator for whether a person is white, black, or another race (`i.race`), an indicator for whether a person's parents live with her or him (`i.parents_r`), and a categorical education level (`i.educ_r`). Because we want to estimate the probability of response given that a unit is in the sample, the survey weights are not used in the regression. The weights inflate the sample to the population. That is, a weighted regression would fit an unconditional census model—not one conditional on the units in the initial sample. Although some levels of the categorical variables are not significantly different from zero, we retain them all.

```
. use http://www.stata-press.com/data/svywt/nhis, clear

. logit resp age i.hisp i.race i.parents_r i.educ_r
Iteration 0:   log likelihood = -2420.9745
Iteration 1:   log likelihood = -2379.7413
Iteration 2:   log likelihood =  -2379.593
Iteration 3:   log likelihood =  -2379.593

Logistic regression                             Number of obs   =       3,911
                                                LR chi2(8)      =       82.76
                                                Prob > chi2     =      0.0000
Log likelihood =  -2379.593                     Pseudo R2       =      0.0171

------------------------------------------------------------------------------
        resp |      Coef.   Std. Err.      z    P>|z|     [95% Conf. Interval]
-------------+----------------------------------------------------------------
         age |  -.0126674   .0022052    -5.74   0.000    -.0169896   -.0083453
      2.hisp |   .3060543   .0911932     3.36   0.001     .1273188    .4847898
             |
        race |
          2  |  -.1595515   .0994161    -1.60   0.109    -.3544035    .0353005
          3  |  -.3739457   .1617283    -2.31   0.021    -.6909274    -.056964
             |
  2.parents_r|    .522474   .1102778     4.74   0.000     .3063335    .7386145
             |
       educ_r |
          2  |   .2489401   .0979235     2.54   0.011     .0570135    .4408666
          3  |   .3462311   .0914344     3.79   0.000      .167023    .5254391
          4  |   .2756157   .1418844     1.94   0.052    -.0024726     .553704
             |
       _cons |   .5827413   .1258812     4.63   0.000     .3360187    .8294638
------------------------------------------------------------------------------
```

The predicted values are computed and divided into quintiles using `cut()` in an `egen` statement in code below. The unweighted mean of the predicted values is equal to the unweighted response rate of 69%—a well-known phenomenon in a fitted logistic model that contains an intercept. The counts of persons by quintile class range from 775 to 790; the counts per class are similar, though not exactly the same. We then compute by class the unweighted and survey-weighted class means of the estimated propensities, the unweighted and survey-weighted response rate, and the median response propensity. These choices are all quite similar in this example, and any of them could be used as the common weight adjustment in each class. Note that the weighted mean function, `wtmean()`, is not part of official Stata and must be installed by typing `ssc install _gwtmean`.

```
. predict predp, pr

. summarize predp
    Variable |        Obs        Mean    Std. Dev.         Min         Max
-------------+--------------------------------------------------------------
       predp |      3,911    .6901048    .0673549    .4526496     .818193
. pctile qpreds=predp, nq(5) genp(percent)
. sort predp
. egen pclass=cut(predp), at(0, 0.6306909, 0.6769733, 0.7137994, 0.7516164, 1.0)

. tabulate pclass
      pclass |      Freq.     Percent        Cum.
-------------+-----------------------------------
           0 |        790       20.20       20.20
   .6306909 |        784       20.05       40.25
   .6769733 |        775       19.82       60.06
   .7137994 |        780       19.94       80.01
   .7516164 |        782       19.99      100.00
-------------+-----------------------------------
       Total |      3,911      100.00

. sort pclass
. egen pavg = mean(predp), by(pclass)
. egen pmed = median(predp), by(pclass)
. egen RR = mean(resp), by(pclass)
. egen pavgwtd = wtmean(predp), weight(svywt) by(pclass)
. egen RRwtd = wtmean(resp), weight(svywt) by(pclass)

. table pclass, contents(mean pavg mean pavgwtd mean RR mean RRwtd mean pmed)
--------------------------------------------------------------------------------
      pclass | mean(pavg) mean(pavg~d)   mean(RR)  mean(RRwtd)   mean(pmed)
-------------+------------------------------------------------------------------
           0 |   .5890679     .591958    .5873418     .5901461     .5951289
   .6306909 |   .6561612    .6559756    .6658163     .6819369     .6568276
   .6769733 |   .6966377       .6969    .6941935     .7012339     .6971353
   .7137994 |   .7326617     .732752    .7076923     .7169269     .7330443
   .7516164 |   .7772833    .7777547    .7966752     .8041545     .7752395
--------------------------------------------------------------------------------

. save "nhis.logistnr.dta", replace
```

◼

If the dataset is large enough, there is no reason to restrict the number of classes to five. Because nhis.dta has 3,911 persons, we could easily use 10 classes and have enough units per class to estimate stable NR adjustments.

The sort pclass command sorts the entire dataset. Each of the summary class propensities is properly merged with the records. For example, if the unweighted response rate is used in each class, the adjustment in the first class is $1/0.5873418 =$ 1.703; in the second class, it is $1/0.6658163 = 1.502$. Lastly, we save the resulting

dataset that includes respondents, nonrespondents, and the `pclass` for each case in
`nhis.logistnr.dta`. This dataset will be used in example 5.11 in chapter 5 to illustrate a technique in replicate variance estimation.

**Example 3.3: Assessing model fit.** The goal of a nonresponse adjustment is to
create a pseudoweight that mollifies any biasing effects linked to less than 100% participation. However, the binary regression model chosen to construct the adjustment may
not address bias and only lower precision through increased variability of the weights.
Thus, evaluating the utility of the model or even a set of models is important.

The fit of binary regression models can be assessed in various ways. One, that
can also be used for the machine learning algorithms discussed later in this chapter,
is to fit the model to a randomly selected subset of the full sample (the "training"
dataset) and then evaluate how well it predicts for the "hold-out" or "test" dataset
(see `ex.3.3_assess.fit.do`). In this case, we select a simple random sample without
replacement of $n = 783$ (20%) from the full dataset, `nhis.dta`, and designate it as the
test sample. The remainder $(3911 - 783 = 3128)$ is the training sample. The code is

```
. generate random = runiform()
. sort random
. generate intest = _n <= 783
. generate intrain = _n > 783
.   * save test dataset
. preserve
. keep if intest == 1
. save nhistest, replace
. restore
```

We then use the training subsample and fit the same logistic model as above. Predicted probabilities of response for the test dataset, `nhistest`, are made.

```
.   * extract training dataset and fit logistic model
. keep if intrain == 1
. logit resp age i.hisp i.race i.parents_r i.educ_r
.   * predict for test dataset
. use nhistest, clear
. predict testp, pr
```

To check the accuracy of the predictions, one approach is to use what is called a
"majority vote" method (James et al. 2013). The predicted probabilities of response
are rounded to 0 or 1, and the number of times are counted in which the rounded
predictions agree with the actual response status. As the first table below shows, the
majority vote classification, `testR`, is correct 69.86% of the time. The second table is a
cross-tabulation of `testR` versus `resp`. Cramér's $V$ (a measure of association) is 0.0695,
which is not very high but, nonetheless, typical in these applications.

```
.   * "majority vote" assignment to R or NR for test data
. generate testR = round(testp)
. generate test_accuracy = testR == resp
```

```
. tabulate test_accuracy
```

| test_accuracy | Freq. | Percent | Cum. |
|---|---|---|---|
| 0 | 236 | 30.14 | 30.14 |
| 1 | 547 | 69.86 | 100.00 |
| Total | 783 | 100.00 | |

```
. tabulate testR resp, column V
```

| Key |
|---|
| *frequency* |
| *column percentage* |

| | resp | | |
|---|---|---|---|
| testR | 0 | 1 | Total |
| 0 | 3 | 1 | 4 |
| | 1.26 | 0.18 | 0.51 |
| 1 | 235 | 544 | 779 |
| | 98.74 | 99.82 | 99.49 |
| Total | 238 | 545 | 783 |
| | 100.00 | 100.00 | 100.00 |

Cramér's V = 0.0695

∎

In addition to model fit, D'Agostino (1998) gives a simple method for checking whether respondents and nonrespondents within propensity classes have the same means on covariates and, hence, about the same propensity of responding within each class. If this type of balance is achieved, propensities within a weighting class will be similar and differ from those of other classes. Additional information is provided in Valliant, Dever, and Kreuter (2013, sec. 13.5).

## 3.3 Tree-based algorithms

The binary regression approach in the previous section has some limitations. The primary limitation is that finding the best model for predicting response propensity may be difficult. We used main effects only in the logistic regression example, but a better model may have some complicated interactions. If the response propensity is not modeled correctly, then nonrespondents within cells may not be MAR. Consequently, the NR adjustments will not properly correct NR bias. Three options discussed in this section are classification and regression trees (CART), random forests, and boosting, all of which may produce better models. James et al. (2013) give an excellent introduction to all of these topics.

### 3.3.1 Classification and regression trees

The problem of forming cells for nonresponse adjustment is a type of classification problem in which a probability of response must be estimated for each cell. One useful algorithm for forming cells is CART (Breiman et al. 1984). As in logistic (or probit or complementary log-log) regression, the CART algorithm predicts response based on a set of covariates. In a series of steps, the dataset is successively split into groups by finding the covariate (or recoding of a covariate) that maximizes the log likelihood for the binary variable that defines whether a unit is a respondent. Values of the covariates must be known for every eligible unit in the sample, ER or ENR. At each step, a different covariate is selected or a different coding of a previously used covariate is used. The splitting process terminates when some user-controllable criteria are met. The cells, or terminal nodes, formed by the combinations of covariates used in the last step are the NR adjustment cells. Advantages of CART compared with propensity modeling are the following:

1. Interactions of covariates are handled automatically.

2. The way in which covariates enter the model does not have to be made explicit. Regardless of whether covariates are continuous or categorical, the splitting algorithms handle them the same way.

3. Selection of which covariates and associated interactions to include is done automatically.

4. Variable values, whether categorical or continuous, are combined (grouped) automatically.

There are currently no Stata estimation commands for CART, but there are several packages available in R. Of particular note is `rpart` (Therneau, Atkinson, and Ripley 2017).

### 3.3.2 Random forests

A single regression tree does tend to overfit by creating a complex set of cells that may not transfer well to a new set of data. That is, in the application to nonresponse adjustment, the fitted model may be a poor representation of the underlying response mechanism. Random forests (Breiman 2001) improve the predictions from a single tree by fitting many trees and averaging the results. To understand how random forests work, we first describe what is known as "bootstrap aggregation" or "bagging". The idea behind bagging is to select an srswr of units from the full sample, fit a regression tree to each, record the results for each tree, and then average the results across all the bootstrap samples. The average result will then have lower variance on the predicted values than the result from a single tree.

Random forests refine bagging by making a small adjustment that reduces the correlation between trees. Instead of using all available covariates for each tree, a subsample of the covariates is used. In datasets where there is one dominant predictor, bagged trees will tend to always split on that predictor first. Consequently, predictions from bagged trees will be highly correlated. By selecting a subset of the covariates, random forests "decorrelate" the predictions across trees, producing lower variance average predictions.

There are currently no Stata estimation commands for random forests. In R, there are different implementations of random forests—not all of which work equally well. The R package `randomForest` uses the original recommendations in Breiman (2001). There is some evidence that this version is prone to incorrectly favor some variables in studies where continuous variables are used in combination with categorical ones or when categorical predictors vary in their number of categories (Strobl et al. 2007). An algorithm that has been shown to have better empirical performance is `cforest` (Strobl et al. 2008). The R package `party` (Hothorn et al. 2016) contains an implementation of `cforest`.

### 3.3.3 Boosting

A technique called boosting also generally improves predictions over those of CART. Boosting allows a set of covariates to be specified without defining their functional relationship to a response variable. As with CART and random forests, the algorithm can find complicated interactions that would be hard to identify with standard regression modeling techniques. Bauer and Kohavi (1999) and Friedman, Hastie, and Tibshirani (2000) compared several boosting variants with the CART method and found that all the versions of boosting had better predictive accuracy than CART on several datasets. Schonlau (2005) reviewed boosting and wrote the community-contributed command, `boost`, that implements the multiple additive regression trees (MART) algorithm described by Friedman, Hastie, and Tibshirani (2000) and Friedman (2001). Schonlau's command can be downloaded by typing `net from http://www.schonlau.net/stata` and then selecting `boost` from among several ado-files that Schonlau has written. Other methods such as Bayesian additive regression trees (BART; Chipman, George, and McCulloch 2010) exist but are not discussed further.

The following is a rough explanation of how boosting works: A series of regression trees is fit with gradually increasing complexity. First, a subset of the full dataset is selected for fitting the model (the "training" set). The initial prediction for each $y_i$ (or in our case, $R_i$) is their mean; call this the base model. A regression tree of a specified size is then fit to the residuals from the simple mean model. When modeling whether a unit responds, we use deviance residuals. The resulting tree is then the model found in the first iteration of the process. The model is updated with additional covariates based on what the regression tree finds. A regression tree is then fit to the residuals from that step. A subsample of the training set may be used to fit the model at each iteration. Selecting a subsample at each iteration helps select a model that yields better predictions on new datasets, not used in fitting. This procedure is repeated a large number of times. The final model is used for prediction.

The code below uses the same set of covariates as in the logistic example above to predict response to the income question in `nhis.dta`. The boost algorithm has several controlling parameters that are described in the help file. The options used in the example are

- `distribution()`: Type of model—normal, logistic, poisson
- `trainfraction()`: Fraction of sample used as the training set
- `bag()`: Fraction of training observations to be used to fit an individual tree
- `interaction()`: Maximum number of interactions allowed in the tree
- `shrink()`: Shrinkage parameter to avoid overfitting

In the example below, 80% of the full sample is used for the training set by specifying `trainfraction(0.8)`. Because the plugin uses the first 80%, the file should be sorted in a random order. This is accomplished by generating a random number (`u1`) uniformly distributed between 0 and 1, and sorting the file based on `u1`. The plugin uses the first `trainfraction()` proportion of the observations to fit the model. In the training set, a random sample of 70% of the observations (`bag(0.7)`) is used to fit the individual trees. The size of the regression tree is determined by `interaction()`, set to 7 here, meaning that up to seven-way interactions are allowed. Pseudo-$R^2$s are computed for the training set and the remainder (the "test" dataset).

Shrinkage is intended to reduce the impact of each additional tree in an effort to avoid overfitting. The intuition behind this idea is that it is better to improve a model by taking many small steps rather than a smaller number of large steps. The smaller the value of `shrink()`, the greater the shrinkage; the default value is 0.01. Decreasing the value of the shrinkage will result in a larger variety of models being tried and more iterations being needed for the algorithm to converge. Several combinations of `interaction()` and `shrink()` should be tried to identify one that gives the most accurate predictions. For this example, we tried all combinations of 3, 5, and 7 interactions and shrinkage of 0.1, 0.01, and 0.001. The choice that gave the best predictive accuracy was 7 interactions and shrinkage $= 0.1$. The `seed()` option of `boost` allows you to rerun this example and get the same results as presented here.

Because `boost` does not allow the use of the `i.*` factor-variable notation, dummies are created manually using `generate` as shown below. The full set of code is in `ex.3.6_boostnr.do`, which contains a number of choices of `interaction()` and `shrink()` other than the one shown below.

### Example 3.4: Boosting example using boost procedure.

```
. use http://www.stata-press.com/data/svywt/nhis, clear
.       * randomize order of the file
. set seed 526544083
. generate double u1 = runiform()
. sort u1
. generate hisp1 = (hisp == 1)
```

```
. generate race1 = (race == 1)
. generate race2 = (race == 2)
. generate race3 = (race == 3)
. generate parents_r1 = (parents_r == 1)
. generate educ_r1 = (educ_r == 1)
. generate educ_r2 = (educ_r == 2)
. generate educ_r3 = (educ_r == 3)
. generate educ_r4 = (educ_r == 4)
. boost resp age hisp1 race1 race2 race3 parents_r1 educ_r1 educ_r2 educ_r3
> educ_r4, distribution(logistic) trainfraction(0.8) interaction(7) shrink(0.1)
> predict(boostpI7_S0p1) bag(0.7) seed(528179368) influence
. label variable boostpI7_S0p1 "Boost predictions I=7 S=0.1"
```

◼

Some of the output from the procedure is shown below. Given the parameter inputs, the algorithm determines the best number of iterations that will maximize the likelihood and reports those as `bestiter=15`. The pseudo-$R^2$ for the training dataset is 0.2819111, which is much higher than the 0.0171 in the logistic regression in example 3.2 above. However, the pseudo-$R^2$ for the test dataset is only 0.00630009, which is very low, illustrating that the model does not predict well when used on the hold-out sample. `boost` also produces the percentage of log likelihood explained (labeled as "Influence" in the output) for each input variable for the best number of iterations. In this example, `age` and the first level of `educ` (high school, general education development (GED) degree, or less) are the most important, while the levels of `race` and the last three levels of `educ` are less important.

```
bestiter= 15
Test R2= .00630009
trainn= 3128
Train R2= .2819111
Influence of each variable (Percent):
55.913888 age
6.2221824 hisp1
0 race1
2.3827924 race2
2.7359757 race3
9.8041284 parents_r1
18.818951 educ_r1
2.2569541 educ_r2
1.8651282 educ_r3
0 educ_r4
```

To check the accuracy of the predictions, we again use the majority vote method. As explained in the `boost` help file, the algorithm uses the first `trainfraction()` of the observations, 80% in this example, to fit the model but will make predictions for the entire dataset that is in memory. The predicted probabilities of response are rounded to 0 or 1, and the number of times counted in which the rounded predictions agree with the actual response status. The following is the Stata code and output of tables of counts. The percentage of units where the algorithm correctly predicted response was 69.29%, which is about the same as the overall response rate. Cramér's $V$ for the full set of 3,911 observations is 0.0714. However, $V$ for the test dataset is only 0.0336. Thus, in this example, boosting is worse than logistic regression for predicting response in the test dataset.

```
. generate boost_pred = round(boostp)
. generate accuracy = boost_pred == resp
. tabulate accuracy
```

| accuracy | Freq. | Percent | Cum. |
|---|---|---|---|
| 0 | 1,201 | 30.71 | 30.71 |
| 1 | 2,710 | 69.29 | 100.00 |
| Total | 3,911 | 100.00 | |

```
. tabulate boost_pred resp, column V
```

| Key |
|---|
| frequency |
| column percentage |

| boost_pred | resp 0 | 1 | Total |
|---|---|---|---|
| 0 | 23 | 12 | 35 |
|  | 1.90 | 0.44 | 0.89 |
| 1 | 1,189 | 2,687 | 3,876 |
|  | 98.10 | 99.56 | 99.11 |
| Total | 1,212 | 2,699 | 3,911 |
|  | 100.00 | 100.00 | 100.00 |

Cramér's V =    0.0714

```
. tabulate boost_pred resp if _n > 3128, column V
```

| Key |
|---|
| frequency |
| column percentage |

| boost_pred | resp 0 | 1 | Total |
|---|---|---|---|
| 0 | 4 | 5 | 9 |
|  | 1.69 | 0.91 | 1.15 |
| 1 | 232 | 542 | 774 |
|  | 98.31 | 99.09 | 98.85 |
| Total | 236 | 547 | 783 |
|  | 100.00 | 100.00 | 100.00 |

Cramér's V =    0.0336

The predicted values can be compared with predictions from the logistic regression with main effects shown only in section 3.2. Figure 3.1 is a plot of the logistic predictions versus the boosted predictions for the probability of response. The relationship between the logistic and the boosted predictions is fairly diffuse. The individual predictions do differ from each other, although in both cases the average prediction is close to the overall response rate of 69%.

```
. logit resp age i.hisp i.race i.parents_r i.educ_r
. predict predp, pr
. twoway scatter predp boostpI7_S0p1 || lfit predp boostpI7_S0p1, legend(off)
> ytitle("Logistic predictions")
```

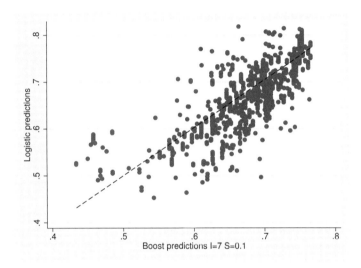

Figure 3.1. Logistic versus boost predictions; reference line is drawn at $y = x$

Another way of visualizing the difference between the logistic and boosted predictions is using boxplots after arraying the predictions into adjustment classes as in section 3.2. Figure 3.2 shows side-by-side boxplots. Both sets of predictions have some extremely low values in the first quintile. This skewness is the reason that a single value is usually used as a nonresponse adjustment in each propensity class. (See `fig.32.logist.boost.boxplot.do` for the commands to create the graph.)

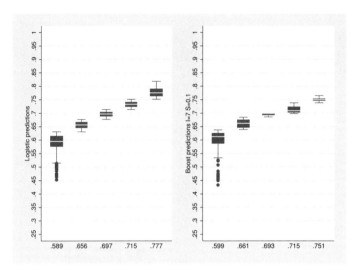

Figure 3.2. Boxplots of logistic and boost predictions.

Choosing one of these options for NR adjustment is a judgment call in any particular survey. In this example, it might be argued that the logistic predicted probabilities would more nearly correct nonresponse bias, but if the model were not completely trusted, the less extreme adjustments from `boost` could be used. In many situations, mistrust of the nonresponse adjustment model may be overriding. We return to the topic of reducing inefficient variation in weights in section 4.5, where unequal weighting effects (UWEs) are discussed.

Once nonresponse adjustments have been made, further improvements can often be made through the calibration techniques covered in the next chapter.

## 3.4    Nonresponse in multistage designs

Before leaving the subject of nonresponse, we must point out a few aspects for multistage designs. Nonresponse can occur at more than one stage of the design depending on the type of unit sampled at the various stages. For example, a geographic area in an area-probability household survey cannot technically refuse to participate; but the person answering the door can refuse for the household. Subsampling within the household and further recruitment cannot occur without details of who lives there. Consequently, multiple adjustments may need to be generated within or across stages of the design to address nonresponse bias.

Keeping with our three-stage household survey example (area, household within area, person within household), we have three base weights:

- $d_{A0} = \pi_i^{-1}$, the area-level (PSU) base weight;
- $d_{H0|A} = \pi_{j|i}^{-1}$, the household-level (second-stage unit) base weight conditioned on the selection of PSU $i$; and
- $d_{P0|A,H} = \pi_{k|i,j}^{-1}$, the person-level (element) base weight conditioned on the selection of second-stage unit $ij$.

The input weight for the household-level nonresponse is the unconditional base weight,

$$d_{H0} = d_{A0}d_{H0|A}$$

adjusted for any household with unknown eligibility. The resulting nonresponse-adjusted base weight (ignoring the eligibility adjustment for convenience) is

$$d_{H1} = d_{H0}a_{H1}$$

where $a_{H1}$ is the household nonresponse adjustment factor calculated most likely with more simplistic techniques such as a weighting class adjustment. This weight is nonzero for eligible households providing information on the number (and potentially the characteristics) of those living in residence, and zero otherwise.

The revised person-level base weight, adjusted for household nonresponse, is defined as

$$d_{H1} = d_{H1}d_{P0|A,H}$$

$$= d_{A0}d_{H0|A}a_{H1}d_{P0|A,H}$$

This weight is used as the input for an unknown eligibility adjustment (if needed) and a subsequent nonresponse adjustment with available area-, household-, and possibly person-level auxiliary information.

Limited or no auxiliary information for a nonresponse adjustment leads analysts to skip this step. Any subsequent weight adjustments, such as calibration (see chapter 4), or model-based analyses are assumed to address the nonresponse bias.

# 4 Calibration and other uses of auxiliary data in weighting

Calibration is usually the last step in weighting and is extremely important in many surveys. Auxiliary data are used to reduce standard errors and to correct coverage problems introduced through the frame or through nonresponse not corrected with a prior adjustment. By auxiliary data, we mean information that is available for the entire frame or target population, either for each individual population unit or in aggregate form. These may be obtainable because a frame of all units in the population is available that has the auxiliary data on each unit. Surveys of business establishments or institutions may have such frames.

The standard application of calibration uses (aggregated) population totals of the auxiliaries to compute weights. Population totals for some variables may be available from a source separate from the survey, like a census. In a business survey, the frame might have the number of employees from an earlier time period for each establishment. In a household survey, counts of persons in groups defined by age, race, and gender may be published from a census, from population projections, or from a separate survey(s). In this chapter, we review several ways to use those data.

Figure 1.1 showed an example of a sample that had both under- and overcoverage. A common situation in household surveys is that some groups of persons are not covered as well as others by the sample design. In the United States, for example, estimates of the numbers of persons from most household surveys are less than counts from the most recent census, with the undercoverage being especially severe for young black or Hispanic males (for example, see U.S. Census Bureau [2014, table 2]). By computing weights that sum to census counts for male blacks and Hispanics, the undercoverage is corrected, at least, in the sense that estimates of population counts match the census counts. Details of some of the techniques for doing this are given in sections 4.1 and 4.3.

Inherent in this discussion is the need for the survey and the control total source(s) to use consistent data collection methods. Both should capture the data in the same way—identical questions and mode of data collection—from the same inferential population. In practice, researchers may have to decide if the level of consistency is close enough. If, however, the frame population differs greatly from the inferential population captured in the control total source(s), then strong assumptions are needed to proceed with weight calibration.

The estimators we cover in this chapter all have a superpopulation model that supports them.[1] Taking a model-based view of inference as discussed in section 1.3, the estimators will be approximately or exactly unbiased for an estimated total of $y$ if the model holds for that $y$ in the population. From a design-based point of view, the estimator will be approximately unbiased in repeated sampling and more efficient than one that does not use the auxiliary data.[2] (Note that if the sample is not a probability sample, as covered in chapter 6, then some type of model-based estimation is the only option for analysis.) We point out the appropriate model for each technique as we go along. Model fitting and checking should be an important step in deciding what form of estimator to use, although practitioners often ignore this step.

## 4.1   Poststratified estimators

Poststratified and raking estimators are two of the most commonly used calibration estimators. They are especially popular in household surveys of persons where the auxiliary variables are indicators for demographic groups. For example, persons may be classified by age group, gender, and race. Poststratification is implemented within calibration weighting classes formed by crossing all categories of the qualitative variables and constructing weights that reproduce the class-specific population counts in the weighted estimates. Poststratification can also be done using a single variable like age group. The poststratified estimator of a total is defined as

$$\widehat{t}_{y\mathrm{PS}} = \sum_{\gamma=1}^{G} N_\gamma \left(\widehat{t}_{y\gamma} \Big/ \widehat{N}_\gamma\right) \tag{4.1}$$

where $\widehat{t}_{y\gamma} = \sum_{s_\gamma} d_i y_i$ is the estimated total of $y$ in weighting class (or poststratum) $\gamma$ based on the input weights, $s_\gamma$ is the set of responding sample units in poststratum $\gamma$, $\widehat{N}_\gamma = \sum_{s_\gamma} d_i$ is the estimated population size of poststratum $\gamma$ based on the input weights, $N_\gamma$ is the population count (also known as a control or control total) for the poststratum $\gamma$, and $G$ is the total number of poststrata. The input weights, $d_i$, can be base weights or weights that have been adjusted for unknown eligibility or nonresponse. The implied final weight for unit $i$ in poststratum $\gamma$ is

$$w_i = d_i \frac{N_\gamma}{\widehat{N}_\gamma} \tag{4.2}$$

where $N_\gamma \big/ \widehat{N}_\gamma$ is the poststratification adjustment (factor). With that definition of the weight, we can write the estimator as $\widehat{t}_{y\mathrm{PS}} = \sum_{i \in s} w_i y_i$, that is, a weighted sum of the data values.

---

1. The term "superpopulation model" means a statistical model that appears to agree with the structure of the data observed in the finite population.
2. This design-based property is contingent on any nonresponse bias having been corrected using the methods in chapter 3. Consequently, a more honest term might be "pseudodesign based", because which units respond is almost never under the survey designer's control.

   The weighting classes are called poststrata because they are applied after the sample is selected and data are collected. This gives flexibility in which variables are used to define the poststrata because they do not have to be available when the sample is designed. Consequently, poststratification is a good way to use auxiliaries that you think are effective predictors of important variables collected in the survey but cannot be easily used for sample selection or for a nonresponse adjustment. For example, in a household survey, many countries do not have a frame of persons that includes race and educational attainment, even though those variables may be correlates of many analysis variables.

   In the model that supports poststratification, each unit in poststratum $\gamma$ has the same mean

$$E_M(y_i) = \mu_\gamma$$

for all units $i$ in group $\gamma$. The units in the sample can be independent under the model, or the model can reflect clustering in the sample (and population) so that elements within clusters are correlated under the model.

**Example 4.1: Poststratification.** The example below reads a sample of 2,000 persons selected from the U.S. National Health Interview (NHIS, `nhis.large.dta`) population in `PracTools`, which has $N = 21588$ persons and poststratifies it using a 5-category age-group variable. The full code is in `ex.4.1_poststrat.do`. The `svyset` statement specifies the design weight as `wt`. Two options are used: `poststrata()`, which names the variable that holds the poststratum values (`age_grp` in this case), and `postweight()`, which gives the variable that contains the poststratification control totals (`poptots`). Each record on the file that is in a given poststratum should have the same value in the `poptots` field. As a check, we run a table of the weighted population estimates for age group. Each estimated count is equal to the control total in each group, plus their standard errors (SEs) are 0. The estimated SEs are 0 for any control total used in a calibration adjustment because the algorithm forces the estimates to be the same in every sample.

```
. use http://www.stata-press.com/data/svywt/nhis_sam.dta
. label define age_lab 1 "<18" 2 "18-24" 3 "25-44" 4 "45-64" 5 "65+"
. label values age_grp age_lab
. svyset [pweight=wt], poststrata(age_grp) postweight(poptots)
        pweight: wt
            VCE: linearized
     Poststrata: age_grp
     Postweight: poptots
    Single unit: missing
       Strata 1: <one>
           SU 1: <observations>
          FPC 1: <zero>
. svy: tabulate age_grp, count se
(running tabulate on estimation sample)
```

```
Number of strata   =        1          Number of obs      =      2,000
Number of PSUs     =    2,000          Population size    =     21,588
N. of poststrata   =        5          Design df          =      1,999
```

| age.grp | count   | se |
|--------:|--------:|---:|
| <18     | 5991    | 0  |
| 18-24   | 2014    | 0  |
| 25-44   | 6124    | 0  |
| 45-64   | 5011    | 0  |
| 65+     | 2448    | 0  |
| Total   | 2.2e+04 |    |

```
Key:  count    =  weighted count
      se       =  linearized standard error of weighted count
```

A shortcoming of the Stata code above is that the poststratified weights in (4.2) are not saved as a separate variable and, consequently, cannot be stored for later use. However, using `svycal`, the code below does compute and save the weights. These weights will be saved if the data file is saved. The population controls are stored in a matrix called `poptotals`. Notice that the matrix contains six positions, with the first holding the sum of the age group control totals that are in positions 2–6 corresponding to the five levels of `age_grp`. The first position is the control total for the constant in the poststratification model, which is the population size. The `gen(ps_wt)` option of `svycal` causes the poststratified weights to be saved in `ps_wt`. `svycal` is a general command that will be used for more elaborate applications of calibration in later sections.

```
. matrix poptotals = 21588, 5991, 2014, 6124, 5011, 2448
. matrix colnames poptotals = _cons 1.age_grp 2.age_grp 3.age_grp 4.age_grp
> 5.age_grp
. svycal regress i.age_grp [pw=wt], gen(ps_wt) totals(poptotals)
```

The poststratified weights can be used in analysis, for example, to estimate the proportion of persons who receive Medicaid (a medical assistance program for the poor) in the United States:

```
. generate medicaid1 = abs(medicaid - 2)
. svyset [pweight=wt], poststrata(age_grp) postweight(poptots)

. svy: mean medicaid1
  (output omitted)
```

|          | Mean    | Linearized Std. Err. | [95% Conf. | Interval] |
|----------|---------|----------------------|------------|-----------|
| medicaid1 | .1052065 | .0070797            | .0913219   | .119091   |

■

> **Warning**. Be wary of specifying the survey design in a way that does not recognize that poststratification has been used. For example, if **svyset** is issued as
>
> ```
> svyset [pweight=ps_wt]
> ```
>
> Stata will use the wrong variance estimation formula. SEs of the estimated age group population counts will be nonzero because we do not give ourselves proper "credit" for poststratifying. SEs may be overestimated for variables for which poststratification does result in precision gains.

If you are the database constructor and are supplying a file with poststratified weights to other users, they need to know both definitions of the poststrata and their control totals to correctly compute SEs. This proviso applies to all other calibrated weights, including raking and general regression that will be covered later in this chapter.

## Recovering control totals from public-use files

Public-use files are often supplied with weights computed via poststratification or another calibration procedure. To correctly compute SEs, Stata needs to be informed that poststratification was used. This requires knowing both the variable(s) used to define poststrata and the population (control) totals. As a data user, you may have to rely on the survey documentation to try to reconstruct both. For example, suppose the documentation says that poststrata were based on age × gender. A survey-weighted tabulation of the estimated population counts in the age × gender table will reproduce the population controls. If the documentation does not clearly list the age categories, you may have to contact the organization that provided the database or make an educated guess as to what they were.

Given your own construction of a poststratum variable and the recovered population totals, svyset can be specified as in example 4.1. Expending this effort is worthwhile to get more honest SEs. These better SEs could well be smaller than if you treated the weights as inverse selection probability weights like in the warning above.

**Example 4.2: Recovering poststratification totals.** We use the same sample as in example 4.1 but now use pswt in the dataset as the weight. (The code is in ex.4.2_recover.poptots.do.) If the poststratified weights, pswt, are treated as inverse selection probabilities, and the total number of persons who are not covered by any kind of medical insurance is estimated, we get these results:

```
. use http://www.stata-press.com/data/svywt/nhis_sam, clear
. label define age_lab 1 "<18" 2 "18-24" 3 "25-44" 4 "45-64" 5 "65+"
. label values age_grp age_lab
. generate notcov1 = abs(notcov - 2)
. svyset [pweight = pswt]

. svy: total notcov1
  (output omitted)
```

|          | Total | Linearized Std. Err. | [95% Conf. Interval] |          |
|---------:|------:|---------------------:|---------------------:|---------:|
| notcov1  | 4045.863 | 195.5004 | 3662.454 | 4429.273 |

```
. svy: total notcov1, over(age_grp)
  (output omitted)
    _subpop_1: age_grp = <18
    _subpop_2: age_grp = 18-24
    _subpop_3: age_grp = 25-44
    _subpop_4: age_grp = 45-64
    _subpop_5: age_grp = 65+
```

| Over | Total | Linearized Std. Err. | [95% Conf. Interval] | |
|-----:|------:|---------------------:|---------------------:|---------:|
| notcov1 | | | | |
| _subpop_1 | 875.2284 | 102.0084 | 675.1727 | 1075.284 |
| _subpop_2 | 689.9815 | 112.391 | 469.5637 | 910.3993 |
| _subpop_3 | 1653.662 | 116.8108 | 1424.576 | 1882.747 |
| _subpop_4 | 775.1823 | 82.7784 | 612.8399 | 937.5248 |
| _subpop_5 | 51.80952 | 25.88503 | 1.044608 | 102.5744 |

Next, suppose that the weights on the file were created by poststratifying by the five age groups. We can recover the population totals with a tabulation.

```
. svy: tabulate age_grp, count format(%12.0f)
```
*(output omitted)*

| age.grp | count |
|---:|---:|
| <18 | 5991 |
| 18-24 | 2014 |
| 25-44 | 6124 |
| 45-64 | 5011 |
| 65+ | 2448 |
| Total | 21588 |

Key:  count    = weighted count

Then, the poststratification control counts are appended to the dataset with the following:

```
. generate pstot = 5991
. replace pstot = 2014 if age_grp == 2
. replace pstot = 6124 if age_grp == 3
. replace pstot =  5011 if age_grp == 4
. replace pstot =  2448 if age_grp == 5
```

Retabulating the estimated total of persons not covered by medical insurance and accounting for the poststratification gives

```
. svyset [pweight=wt], poststrata(age_grp) postweight(pstot)

.      * nocov1 totals overall and by age_grp
. svy: total notcov1
```
*(output omitted)*

| | Total | Linearized Std. Err. | [95% Conf. Interval] | |
|---:|---:|---:|---:|---:|
| notcov1 | 4100.936 | 190.2382 | 3727.847 | 4474.025 |

```
. svy: total notcov1, over(age_grp)
```
*(output omitted)*

| Over | Total | Linearized Std. Err. | [95% Conf. Interval] | |
|---|---:|---:|---:|---:|
| notcov1 | | | | |
| _subpop_1 | 882.4917 | 96.73621 | 692.7757 | 1072.208 |
| _subpop_2 | 703 | 93.26878 | 520.0842 | 885.9158 |
| _subpop_3 | 1671.016 | 105.6485 | 1463.822 | 1878.211 |
| _subpop_4 | 791.2105 | 79.24024 | 635.807 | 946.614 |
| _subpop_5 | 53.21739 | 26.32456 | 1.590482 | 104.8443 |

Note that we do not have to compute new poststratified weights using `svycal` as was done in example 4.1 because `wt` is already poststratified. The SE on the full population estimate is reduced by about 2.7% $(1 - 190.2382/195.5004)$ when the poststratification is accounted for. There are also reductions in the SEs of estimated totals for individual age groups. The reason that SEs are smaller when poststratification is properly credited is that there are substantial differences among age groups in the proportions not covered by medical insurance as is evident from the table below. In other words, there is an association between the poststrata and the variable of interest.

```
. svy: mean notcov1, over(age_grp)
  (output omitted)
```

|  | | Mean | Linearized Std. Err. | [95% Conf. Interval] | |
|---|---|---|---|---|---|
| Over | | | | | |
| notcov1 | | | | | |
| _subpop_1 | | .1473029 | .0161469 | .1156361 | .1789697 |
| _subpop_2 | | .3490566 | .0463102 | .2582344 | .4398788 |
| _subpop_3 | | .2728636 | .0172515 | .2390304 | .3066968 |
| _subpop_4 | | .1578947 | .0158133 | .1268823 | .1889072 |
| _subpop_5 | | .0217391 | .0107535 | .0006497 | .0428286 |

Finally, we should note that any reduction in SEs will be most important for estimated totals. Poststratification may not improve the precision of estimated proportions (or any ratio estimator where weights are included in the numerator and denominator) much.                                                                              ∎

Control totals, however, are not always exactly recoverable from public-use files. As discussed in Kim, Li, and Valliant (2007) some weighting cells, but not an entire level of a complicated set of poststrata, may need to be collapsed because of small or zero sample size. For example, say that poststrata are constructed by gender × race (Hispanic, non-Hispanic [NH], white, NH black, NH Asian, and NH other). If there are few respondents in the female NH Asian weighting cell (say, 10) but plenty of NH Asian males, then the statistician may choose to collapse NH Asian females with NH other females but leave the category alone for males. The associated documentation may not include such details. Therefore, we recommend checking the respondent sample size for the poststrata; if counts are small, then producing population control totals for your own version of collapsed cells or for a raking estimator (discussed in section 4.2) may be preferred.

## 4.2  Raking estimators

Raking is another commonly used method of adjusting weights to control totals. In this method, marginal population controls can be used for two or more variables. Raking is often used when several variables are predictive of either coverage or the analysis variables (or both) but the sample sizes in some cells would be small if all variables were fully crossed as they would be for poststratification based on multiple variables. Another reason to use raking is when control totals are only available at the margins, say, in a published report but not for the corresponding poststrata. Like poststratification, raking has an associated linear model. Taking the case of two raking dimensions, the model mean for unit $i$ in level $j$ of the first variable and level $k$ of the second is

$$E_M(y_i) = \mu + \alpha_j + \beta_k$$

where $\alpha_j$ and $\beta_k$ are main effects. Even with two dimensions several variables can be used because a dimension can be a cross-classification. For example, one dimension might be (age group) × education and the other race × (income group).

**Example 4.3: Raking with svycal.** The code below (also in `ex.4.3_rake.do`) uses `svycal` to rake to control totals for age group and Hispanic using the same `nhis_sam` sample as above. The variable `hisp` is recoded from 4 categories to 3 in `hispr` because the sample size in category 4 (non-Hispanic all other race groups) is small. The `gen(rake_wt)` option saves the weights in a variable called `rake_wt`. The `totals()` option lists the control totals in order for the levels of `age_grp` and `hispr`.

```
. use http://www.stata-press.com/data/svywt/nhis_sam, clear
. label define age_lab 1 "<18" 2 "18-24" 3 "25-44" 4 "45-64" 5 "65+"
. label values age_grp age_lab
. recode hisp (1=1) (2=2) (3=3) (4=3), generate(hispr)
. svycal rake i.age_grp i.hispr [pw=wt], gen(rake_wt)
> totals(_cons=21588 1.age_grp=5991 2.age_grp=2014 3.age_grp=6124
> 4.age_grp=5011 5.age_grp=2448 1.hispr=5031 2.hispr=12637 3.hispr=3920)

. summarize(rake_wt)
```

| Variable | Obs | Mean | Std. Dev. | Min | Max |
|---|---|---|---|---|---|
| rake_wt | 2,000 | 10.794 | 2.478362 | 8.909467 | 20.06054 |

Using the raked weights to estimate the proportion of persons receiving Medicaid, we get

```
. svyset [pweight=wt], rake(i.age_grp i.hispr,
> totals(_cons=21588 1.age_grp=5991 2.age_grp=2014 3.age_grp=6124
> 4.age_grp=5011 5.age_grp=2448 1.hispr=5031 2.hispr=12637 3.hispr=3920))
. generate medicaid1 = abs(medicaid - 2)

. svy: mean medicaid1
  (output omitted )
```

| | Mean | Linearized Std. Err. | [95% Conf. Interval] | |
|---|---|---|---|---|
| medicaid1 | .1067731 | .0070633 | .0929209 | .1206254 |

which is similar to the estimated mean and SE in poststratification where only age group was used. Notice that in the svyset statement above, the specification of totals in svycal includes marginal population counts for both age group and Hispanic.

The rake() option also allows bounds to be set on the relative change of the raked weights to the input weights. This code sets lower and upper bounds on the ratio of rake_wt/wt using the ll() and ul() options:

```
. svycal rake i.age_grp i.hispr [pw=wt], gen(rake_wtB) totals(_cons=21588
> 1.age_grp=5991 2.age_grp=2014 3.age_grp=6124 4.age_grp=5011 5.age_grp=2448
> 1.hispr=5031 2.hispr=12637 3.hispr=3920) ll(0.8) ul(1.2)
```

This option can be used if unbounded raking makes some extremely large adjustments to the input weights that the analyst feels are untrustworthy. The ll() and ul() options can also be used in svycal regress discussed below. A key point to remember is that these options put a bound on the weight adjustments not on the weights themselves. The formal statement of the constraint that is put on the final weights $w_i$ is

$$L \leq w_i/d_i \leq U$$

where $d_i$ is the input weight for unit $i$ and $L$ and $U$ are, respectively, the lower and upper bounds on the weight ratio.

The sizes of $L$ and $U$ will depend in part on the quality of the input weights. $L$ can be set so that no final weight is less than or equal to 0 or allowed to be less than 1 on the premise that each unit should at least represent itself. $U$ will depend on how large the adjustments need to be to compensate for nonresponse and noncoverage. A sample with very low response or poor coverage of the target population will need larger upward adjustments to the input weights than a sample with high response and near complete coverage.

Another versatile command for raking is ipfraking (Kolenikov 2014) (install by typing net install http://www.stata-journal.com/software/sj14-1/st0323). It has several useful features, including weight trimming and diagnostics. The code below repeats the example above, in which the NHIS sample is raked to margins for age groups and Hispanicity. Some preliminary coding is needed in addition to the labeling and recoding of the hisp variable, which was also done above. The vectors of control totals must be stored as matrices with three requirements:

1. Each matrix must be a $1 \times c$ row vector, where $c$ is the number of control totals in the matrix.

2. Each matrix must have column names in Stata estimation results format, that is, *varname*: #.

3. Each matrix must have a row name that contains the categorical variable for which the totals were computed.

Because of requirement 2 above, the command generate _one = 1 is issued below. This creates the variable _one, which is also set as the prefix to the column names in the control matrices (for example, `matrix coleq age_tot = _one`). This puts the column names in estimation results format. The line `matrix list age_tot` shows the result for the `age_tot` matrix.

**Example 4.4: Raking with ipfraking.** The code for this example is in `ex.4.4_ipfraking.do`.

```
. use http://www.stata-press.com/data/svywt/nhis_sam, clear
. label define age_lab 1 "<18" 2 "18-24" 3 "25-44" 4 "45-64" 5 "65+"
. label values age_grp age_lab
. recode hisp (1=1) (2=2) (3=3) (4=3), gen(hispr)
. generate _one = 1
. matrix age_tot = (5991, 2014, 6124, 5011, 2448)
. matrix coleq age_tot = _one
. matrix rownames age_tot = age_grp
. matrix colnames age_tot = 1 2 3 4 5

. matrix list age_tot
age_tot[1,5]
          _one:  _one:  _one:  _one:  _one:
             1      2      3      4      5
age_grp   5991   2014   6124   5011   2448

. matrix hispr_tot = (5031, 12637, 3920)
. matrix coleq hispr_tot = _one
. matrix rownames hispr_tot = hispr
. matrix colnames hispr_tot = 1 2 3

. ipfraking [pw=wt], ctotal(age_tot hispr_tot) generate(rakedwgt1)
Iteration 1, max rel difference of raked weights = .12813266
Iteration 2, max rel difference of raked weights = .00820574
Iteration 3, max rel difference of raked weights = .00028074
Iteration 4, max rel difference of raked weights = .00001263
Iteration 5, max rel difference of raked weights = 5.767e-07
    Summary of the weight changes
                 |   Mean    Std. dev.    Min      Max       CV
    -------------+-------------------------------------------------
    Orig weights |  10.792    2.3236    9.2546    18.424    .2153
    Raked weights|  10.794    2.4784    8.9095    20.061    .2296
    Adjust factor|  0.9981              0.9627    1.1372
```

Notice that when the `ipfraking` command is issued, it does not have a separate varlist of the calibrating variables. The procedure understands what fields in the data file are the calibration variables because the row names of the control matrices must match exactly the names of those fields. `ipfraking` also assumes that each calibrating variable is categorical, which is a requirement that is not needed for calibration in general. Consequently, we do not have to create dummy (0–1) variables based on `age_grp` and `hispr` or use the `i.*` notation as in `svycal`.

Convergence occurs when the maximum relative change in any weight is less than a tolerance, which, by default, is set at $10^{-6}$. As seen above, five iterations were required. A summary of the original and raked weights is printed by default, along with histograms of the raked weights and adjustment factors (not shown here). The `ipfraking` command also allows lower and upper bounds to be placed on the weights themselves or on the factor by which weights change because of raking. Note that if the lower and upper bounds are too restrictive, then the algorithm may not converge at all.

A third command for raking is `sreweight` (Pacifico 2014), which will also perform the more general calibration described in section 4.3. This package can be installed with `net install http://www.stata-journal.com/software/sj14-1/st0322`. The setup is somewhat different from `svycal` and `ipfraking`, but for the NHIS example, the results are the same. The control totals must all be in one column matrix, `alltots`, shown below. Indicator variables must also be created because `sreweight` can use either categorical or continuous auxiliaries. In the call to `sreweight`, both the calibration variables, `age1-age5 hispr1-hispr3`, and the control totals, `alltots`, must be listed. The parameter `sweight` is the input weight, whereas `nweight` is the raked weight. The parameter `df` allows one of six different distance functions to be specified (see the help file or Pacifico [2014] for details).

**Example 4.5: Raking with sreweight.** The full code is in `ex.4.5_sreweight.do`.

```
. matrix alltots = (5991 \ 2014 \ 6124 \ 5011 \ 2448 \ 5031 \ 12637 \ 3920)
.     * Create dummy vars for age_grp
. generate age1 = (age_grp == 1)
. generate age2 = (age_grp == 2)
. generate age3 = (age_grp == 3)
. generate age4 = (age_grp == 4)
. generate age5 = (age_grp == 5)
. generate hispr1 = (hispr == 1)
. generate hispr2 = (hispr == 2)
. generate hispr3 = (hispr == 3)
```

```
. sreweight age1-age5 hispr1-hispr3, sweight(wt) nweight(rakedwgt2)
> total(alltots) df(c)
Iteration 1
Iteration 2
Iteration 3 - Converged

Survey and calibrated totals
```

| Variable | Original | New |
|---|---|---|
| age1 | 5997 | 5991 |
| age2 | 1990 | 2014 |
| age3 | 6238 | 6124 |
| age4 | 5025 | 5011 |
| age5 | 2335 | 2448 |
| hispr1 | 5078 | 5031 |
| hispr2 | 12858 | 12637 |
| hispr3 | 3648 | 3920 |

```
Note: type-c distance function used
```

This procedure uses a somewhat different criterion for convergence based on the difference between the controls and their estimates and on the difference in a distance function between iterations. As shown above, convergence based on those criteria is reached in three steps.

A shortcoming of `ipfraking` and `sreweight` is that neither provides a convenient way of estimating SEs (unlike `svycal`). One option would be to run `ipfraking` or `sreweight` on each of a set of replicates and then use replication variance estimates, which we discuss later in section 5.4. The `svr` package (Winter 2002) supports several replication methods and is covered in more detail in the later section. As illustrated in example 5.4, the `survwgt` command in the `svr` package will also do raking. We delay showing the details until section 5.4, after we explain replication.

## 4.3  More general calibration estimation

A method of calibration that is somewhat more general than poststratification or raking is general regression (GREG; Särndal 2007) estimation. A GREG can use both qualitative and quantitative covariates. Suppose that the sample size is $n$. The GREG estimator of a total for $y$ is

$$
\begin{aligned}
\widehat{t}_{y\text{GREG}} &= \widehat{t}_y + \left(\mathbf{t}_x - \widehat{\mathbf{t}}_x\right)^T \widehat{\mathbf{B}} \\
&= \sum_{i \in s} \left\{ 1 + \left(\mathbf{t}_x - \widehat{\mathbf{t}}_x\right)^T \left(\mathbf{X}^T \mathbf{D} \mathbf{V}^{-1} \mathbf{X}\right)^{-1} \mathbf{x}_i / v_i \right\} d_i y_i
\end{aligned}
\tag{4.3}
$$

where $\widehat{t}_y = \sum_s d_i y_i$ is the estimator of the total based on the $d_i$ input weights, the superscript $T$ represents the transpose of a vector or matrix, $\mathbf{t}_x = (t_{x1}, \ldots, t_{xp})^T$ is the

$p \times 1$ vector of population (control) totals of the $p$ auxiliaries using the number of rows by the number of columns matrix notation, $\widehat{\mathbf{t}}_x = \sum_s d_i \mathbf{x}_i$ is the estimate of totals of the $x$'s based on the $d_i$ weights, $\mathbf{x}_i$ is the $p \times 1$ vector of auxiliary values for the $i$th sample unit,

$$\mathbf{D} = \operatorname{diag}(d_i) \text{ is the } n \times n \text{ diagonal matrix of input weights}$$

$$\mathbf{X} = \begin{pmatrix} \mathbf{x}_1^T \\ \mathbf{x}_2^T \\ \vdots \\ \mathbf{x}_n^T \end{pmatrix} \text{ is the } n \times p \text{ matrix of auxiliaries for the } n \text{ sample units}$$

$$\widehat{\mathbf{B}} = \left( \mathbf{X}^T \mathbf{D} \mathbf{V}^{-1} \mathbf{X} \right)^{-1} \mathbf{X}^T \mathbf{D} \mathbf{V}^{-1} \mathbf{y}$$

with $\mathbf{y} = (y_1, \ldots, y_n)^T$ being the vector of $y$'s for the sample units and $\mathbf{V} = \operatorname{diag}(v_i)$ being an $n \times n$ diagonal matrix of values associated with the variance parameters in an underlying linear model. It is possible to formulate the GREG using a block-diagonal or some other nondiagonal covariance matrix, but this is seldom done in practice.

The $p \times 1$ vector, $\widehat{\mathbf{B}}$, is an estimator of the slope vector in the model $y_i = \mathbf{x}_i^T \beta + \varepsilon_i$, where the $\varepsilon_i$ have mean 0 and variance $v_i$. Note that in the case of srswor design and base weights, $\widehat{\mathbf{B}}$ reduces to $\left( \mathbf{X}^T \mathbf{X} \right)^{-1} \mathbf{X}^T \mathbf{y}$, the ordinary least squares estimator. If the model errors were all 0, then $\widehat{\mathbf{B}} = \beta$ and the GREG reduces to $\mathbf{t}_x^T \beta$, which is also the population sum of the $y$'s, $t_y$. In that case, the $y$ for each unit in the population can be predicted without error as $\mathbf{x}_i^T \beta$, and the GREG would be exactly equal to $t_y$ in every sample. As a result, the better the predictor that $x$ is of $y$, the smaller the variance of the GREG.

An estimated total for $y$ is calculated as $\widehat{t}_{y\mathrm{GREG}} = \sum_s w_i y_i$, a function of the weights resulting from the calibration procedure. The weights have the form

$$w_i = d_i g_i$$

where $g_i$ is the term in braces in (4.3). The weights, $w_i$, are also the solution to a minimization problem in which a squared distance between the initial weights and the final weights is minimized. The term $g_i$ is called the $g$ weight or a calibration adjustment (factor) (Särndal, Swensson, and Wretman 1992). Note that the final $w_i$ weights do not depend on any analysis variables ($y$'s). As a result, the same set of weights can be used for any estimated total. Using the GREG weights, a (ratio) mean would be estimated as $\widehat{\bar{y}}_{\mathrm{GREG}} = \sum_s w_i y_i / \sum_s w_i$, which is the standard approach in Stata.

We demonstrate calibration with the `svycal` function below. In a real application, modeling of important analysis variables should be done to identify covariates from among the set of corresponding population totals. Correcting for coverage and nonresponse problems should also be considered when selecting covariates. We bypass those important steps here and mainly show the `svycal` syntax.

**Example 4.6: GREG with quantitative and qualitative covariates.** To illustrate the calculation of GREG weights, we use an stsrs sample of 120 hospitals from the Survey of Mental Health Organizations (SMHO) in the `PracTools` R package. The code for this and the next example is in `ex.4.6_4.7_smho.greg.do`. The sample is stratified by type of hospital (`hosptype`); in each stratum, an srswor of 30 hospitals was selected. `svycal` is used to compute weights based on the number of patients on the hospital's role at the end of the year (`EOYCNT`) and the number of beds within the hospital type. The term `i.hosptype#c.BEDS` specifies an interaction between hospital type, used as a categorical variable, and the continuous variable `BEDS`. The `c.` prefix on `BEDS` ensures that it is treated as continuous so that a separate slope is fit for beds within each hospital type. The `gen(gregwt1)` option causes the calibrated weights to be saved as `gregwt1`.

The `totals()` option of `svycal regress` below follows the same syntax as Stata's `from()` option for starting values in maximization problems (see `help maximize` from within Stata). The `copy` suboption tells the internal parser that it is getting a list of numbers to be converted into a vector with the column names derived from the specified model. It is critical to list the values of the totals in the same order as they appear in the calibration model. In this case, the `noconstant` option (no constant) is included in `totals` so that the total number of hospitals is not included in the control totals; otherwise, an intercept is included in the model by default.

The estimated totals of beds within hospital types have zero SEs as shown in the output of `svy: total BEDS, over(hosptype)`. The same is true for the estimated total of `EOYCNT` (not shown here). This is as it should be because the calibration model includes a different slope on beds within each hospital type.

```
. use http://www.stata-press.com/data/svywt/smhosam, clear

. svycal regress EOYCNT i.hosptype#c.BEDS [pw = wt], gen(gregwt1)
> totals(505345 37978 13066 9573 10077, copy) noconstant

. svyset ID [pweight = gregwt1], strata(hosptype)
> regress(EOYCNT i.hosptype#c.BEDS, totals(505345 37978 13066 9573 10077, copy)
> noconstant)
```

```
. svy: total BEDS, over(hosptype)
(running total on estimation sample)

Survey: Total estimation

Number of strata =       4          Number of obs    =         120
Number of PSUs   =     120          Population size = 889.062959
Calibration      : regress          Design df        =         116
                 1: hosptype = 1
                 2: hosptype = 2
                 3: hosptype = 3
                 5: hosptype = 5
```

|                | | Linearized | | |
|---|---|---|---|---|
| Over | Total | Std. Err. | [95% Conf. Interval] | |
| BEDS | | | | |
| 1 | 37978 | 9.13e-13 | 37978 | 37978 |
| 2 | 13066 | 3.47e-14 | 13066 | 13066 |
| 3 | 9573 | 2.04e-15 | 9573 | 9573 |
| 5 | 10077 | 4.64e-13 | 10077 | 10077 |

The estimated mean of expenditures per hospital (in millions) is 10.36 as shown below with an SE of 0.8577 and a coefficient of variation (cv) of 8.28%.[3]

```
. generate exptot = EXPTOTAL/10^6

. svy: mean exptot

(output omitted)
```

|        | | Linearized | | |
|---|---|---|---|---|
|        | Mean | Std. Err. | [95% Conf. Interval] | |
| exptot | 10.35802 | .8577431 | 8.659149 | 12.05689 |

```
. estat cv
```

|        | | Linearized | |
|---|---|---|---|
|        | Mean | Std. Err. | CV (%) |
| exptot | 10.35802 | .8577431 | 8.28096 |

For comparison with the GREG results, we compute the estimated mean expenditures per hospital using the stratified design and its inverse selection probability weight wt. The estimator in that case is $\widehat{\bar{y}} = \sum_h \sum_{i \in s_h} w_{hi} y_{hi} / \sum_h \sum_{i \in s_h} w_{hi} = N^{-1} \sum_h N_h \bar{y}_h$ because $w_{hi} = N_h/n_h$. The SE is about 0.9491, and the cv of the mean is 9.6%. The SE from the GREG estimator is about 20% lower $(1 - 0.8577431/0.9491147)$ than that of the stsrs estimator, $\widehat{\bar{y}}$.

---

3. The cv is a measure of relative precision that is comparable across all types of variables and is calculated as $100 \times$ SE(estimate) / estimate.

```
. svyset ID [pweight = wt], strata(hosptype)

. svy: mean exptot
  (output omitted)
```

|        | Mean    | Linearized Std. Err. | [95% Conf. Interval] |          |
|--------|---------|----------------------|----------------------|----------|
| exptot | 9.88704 | .9491147             | 8.007199             | 11.76688 |

```
. estat cv
```

|        | Mean    | Linearized Std. Err. | CV (%)  |
|--------|---------|----------------------|---------|
| exptot | 9.88704 | .9491147             | 9.59958 |

■

To have any gains from calibration reflected in SE estimates, it is important to correctly specify the options in `svyset` as illustrated in the next example, which uses the GREG weights from example 4.6.

**Example 4.7: GREG weights with incorrect specification of svyset.** If the `regress` option is omitted from `svyset`, Stata will not realize that the weights are calibrated. The weights will be treated as inverse selection probabilities and the with-replacement variance estimator, discussed in section 5.2, will be used. As shown below, this would be a serious mistake because the SE of estimated mean expenditures per hospital is 10.1%—not the correct 8.28% in example 4.6.

```
. svyset ID [pweight = gregwt1], strata(hosptype)
. svy: mean exptot

. estat cv
```

|        | Mean     | Linearized Std. Err. | CV (%)  |
|--------|----------|----------------------|---------|
| exptot | 10.35802 | 1.04789              | 10.1167 |

■

## GREG with bounded weight changes

A problem with GREG estimation is that it can produce negative weights. This is disturbing both intuitively (how can a unit represent a negative number of units) and practically (some software packages, including Stata, do not allow negative weights). To combat this, `svycal` has an option that allows the relative change in the input weights to be bounded. The method of bounding was described in section 4.2 in the context of raking. If the initial weights are positive, the final calibrated weights will also be positive. The next example uses a sample from the SMHO population that was selected with probability proportional to number of beds in each facility. Hospitals with five or fewer inpatient beds in the population file were recoded to have five beds before the sample was selected.

First, we calibrate to the totals of `SEENCNT`, `EOYCNT`, and the count of beds separately for four hospital types. This is the same calibration model as in example 4.6 except that `SEENCNT` is added. The input weights, `d`, which are inverses of the pps selection probabilities, range from 2.71 to 33.68. But, after calibration with no bounding, the range of `gregwt` is $-0.30$ to 33.14. Stata does warn you about this and suggests using the `ll()` option as shown in the note that follows `svycal`.

**Example 4.8: GREG can create negative weights.** The code for this and the next example is in `ex.4.8_4.9_smho.greg.bdd.do`.

```
. use http://www.stata-press.com/data/svywt/smhosam.pps.dta, clear

. svycal regress SEENCNT EOYCNT i.hosptype#c.BEDS [pw = d], gen(gregwt)
> totals(1349241 505345 37978 13066 9573 10077, copy) noconstant
note: Negative weights resulted from calibration.  Consider specifying ll()
      option to eliminate negative weights.
. summarize d
    Variable |        Obs        Mean    Std. Dev.        Min        Max
-------------+--------------------------------------------------------------
           d |         80    8.762516     4.88059   2.713943   33.67915
. summarize gregwt
    Variable |        Obs        Mean    Std. Dev.        Min        Max
-------------+--------------------------------------------------------------
      gregwt |         80    8.979236    4.895928  -.3037212   33.14324
```

If we try to use these `gregwts` for analysis, Stata reports that negative weights were encountered and produces no output.

```
. svyset ID [pweight = gregwt], strata(hosptype) regress(SEENCNT EOYCNT
> i.hosptype#c.BEDS, totals(1349241 505345 37978 13066 9573 10077, copy)
> noconstant)

. svy: total BEDS, over(hosptype)
(running total on estimation sample)
negative weights encountered
r(402);
```

◼

Next, we can bound the ratio of the GREG weights to the input weights with the `ll()` and `ul()` options. In this case, the ratio of the GREG to the input weight is restricted to be between 0.4 and 3.0. The range of the resulting weights is 1.37 to 33.03.

**Example 4.9: GREG with bounded weight changes using svycal.**

```
. svycal regress SEENCNT EOYCNT i.hosptype#c.BEDS [pw = d], gen(gregwtB)
> totals(1349241 505345 37978 13066 9573 10077, copy) noconstant ll(0.4) ul(3.0)

. svyset ID [pweight = gregwtB], strata(hosptype) regress(SEENCNT EOYCNT
> i.hosptype#c.BEDS, totals(1349241 505345 37978 13066 9573 10077, copy)
> noconstant)

. summarize gregwtB
```

| Variable | Obs | Mean | Std. Dev. | Min | Max |
|---|---|---|---|---|---|
| gregwtB | 80 | 8.956736 | 4.864207 | 1.374342 | 33.03271 |

∎

Figure 4.1 plots the unbounded versus bounded GREG weights for this example. Bounding leaves most weights about the same but does move the $-0.30$ weight to 1.37 and modifies weights for a few other cases so that the controls can still be satisfied. Generally, the problem of negative weights will not occur if the input weights are far from zero.

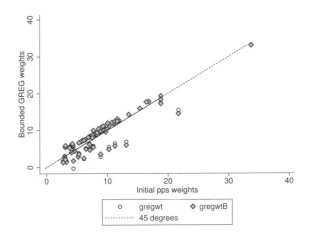

Figure 4.1. Negative GREG weights corrected by weight bounding

## 4.4   Calibration to sample estimates

In some surveys, population counts or totals are not available for the auxiliaries that seem useful for calibration. In those cases, estimates from other surveys may be used as control totals. For example, in household surveys, control totals based on the most recent census in a country may be out of date and the best counts are from recent surveys. Some countries conduct a population census every 10 years that requires field enumeration. Other countries (for example, Denmark, Iceland, Sweden) rely on registers. Some countries like the Netherlands and Switzerland use a combination of registers and surveys to update their population counts. Other countries do not have full population censuses. Germany, for example, has not conducted one since 1987.

Dever and Valliant (2010) review some surveys that use estimated control (ECs) totals. For example, Nadimpalli, Judkins, and Chu (2004) adjusted weights for the 2003 National Survey of Parents and Youth to the number of United States households with children ages 9–18 estimated from the Current Population Survey (CPS) using a ratio-raking algorithm. Nontraditional controls may also be used, particularly in nonprobability surveys (see chapter 6). These may be estimates for quantities that are felt to be predictive of things being measured in a particular survey but are not the standard choices of demographics like age, sex, etc. Schonlau, van Soest, and Kapteyn (2007), for instance, studied whether so-called "webographic" variables, like whether airport searches were perceived as a privacy violation or whether persons allowed websites to store their credit card information, could be used as covariates for estimation in a web survey. The control totals for such specialized covariates would come from a survey that is separate from the one being weighted (that is, the analytic survey). The survey that provides the ECs is usually called the benchmark survey.

Approximately unbiased estimated controls (or statistically consistent) serve the same purposes as fixed controls do: correction for undercoverage and nonresponse bias and reduction of standard errors. To create weights, the estimated controls are treated exactly the same as fixed controls. Consequently, the weighting methods of poststratification, raking, and GREG are applied as described in earlier sections.

The technical complication that comes from using ECs is that they do contribute to the variance of the survey estimates. If the survey that provides the ECs is substantially larger than the analytic survey, then this extra variance contribution associated with the control total estimation can be ignored. In the United States, one source of ECs is the American Community Survey (ACS), which has an annual sample size of about 3.5 million addresses.[4] The size of the ACS dwarfs any survey that might use ACS estimates as controls. Consequently, treating the ECs as fixed is reasonable. However, if the survey providing the ECs is near the size of the analytic survey, ignoring the imprecision of the ECs would be an error. The general variance formulas that account for variance of the ECs are given in Dever and Valliant (2010, eq. 5) for poststratification and in Dever and Valliant (2016, eq. 12) for GREG estimation. These general formulas are complicated, but when the analytic and benchmark surveys can be treated as statistically independent, the variance formula boils down to the variance treating the controls as fixed plus a term that depends on the covariance matrix of the ECs.

---

4. See ACS Design and Methodology report
   https://www.census.gov/programs-surveys/acs/methodology/design-and-methodology.html.

To illustrate, take the poststratified estimator of a total in (4.1). When $N_\gamma$ is estimated from a benchmark survey by $\widehat{N}_{B\gamma}$, the EC poststratified estimator is

$$\widehat{t}_{yPS} = \sum_{\gamma=1}^{G} \widehat{N}_{B\gamma} \left(\widehat{t}_{y\gamma} \big/ \widehat{N}_\gamma\right)$$

If the ECs are uncorrelated with each other, the estimate of the extra contribution to the variance of a poststratified total is

$$\sum_{\gamma=1}^{G} \widehat{\overline{y}}_\gamma^2 \; v_B\left(\widehat{N}_{B\gamma}\right) \tag{4.4}$$

where $\widehat{\overline{y}}_\gamma = \widehat{t}_{y\gamma}/\widehat{N}_\gamma$ in (4.1) and $v_B(\widehat{N}_{B\gamma})$ is an estimate of the variance of the EC population count in poststratum $\gamma$. This term treats the means from the analytic survey, $\widehat{\overline{y}}_\gamma$, as fixed. The variance estimate, $v_B(\widehat{N}_{B\gamma})$, would typically come from the benchmark survey. The problem with (4.4) is that it is often too small, as shown in Dever and Valliant (2010), because it ignores covariances among $\widehat{N}_{B1}, \ldots \widehat{N}_{BG}$. Alternatively, there are some quite complex replication variances in Dever and Valliant (2010, 2016) that do account for the covariances. None of the options for accounting for estimated controls are available in Stata (or any other off-the-shelf software packages) and will require special programming.

## 4.5   Weight variability

The various adjustments covered earlier—unknown eligibility, nonresponse adjustment, and calibration—will typically increase the variability of weights. Ideally, the weight adjustments will remove bias, adjust for coverage errors, and reduce variances—effectively lowering the mean square error (= bias squared + variance). However, if the adjustments do not accomplish those purposes, the increased weight variability is detrimental.

An oft-used measure of weight variability, popularized by Kish (1965), is

$$\text{deff}_K(w) = 1 + \text{relvar}(w) \tag{4.5}$$
$$\equiv 1 + n^{-1} \sum_s (w_i - \overline{w})^2 / \overline{w}^2$$

where $\text{relvar}(w) = n^{-1} \sum_s (w_i - \overline{w})^2 / \overline{w}^2$, the relative variance of the input weights, with $\overline{w} = n^{-1} \sum_s w_i$. This measure is also referred to as an unequal weighting effect (UWE) to distinguish this calculation from one that involves the $y$ variables. The larger $\text{deff}_K$ is, the more inefficient the weights are feared to be. Expression (4.5) is derived as a special case of the design effect of a stratified simple random sample mean, $V(\overline{y}_{st})/V_{srs}(\overline{y})$. If the stratum population variances are all equal, then a proportional allocation with $n_h/N_h = n/N$ is optimal. In that case, the weight for each sample case is the same,

$N_h/n_h = N/n$. Thus, in that restrictive case, $\text{deff}_K$ is a measure of the inefficiency of the weighting system compared with an optimal, equally weighted sample. However, $\text{deff}_K$ is sometimes used indiscriminately for any design—not just stsrs—and in cases where varying weights are actually efficient.

In example 4.6, the GREG weighting reduced the SE of the mean expenditures (in millions) from 0.949 with the stsrs input weights to 0.858. As shown in example 4.10 below, $\text{deff}_K = 1.211$ for the GREG weights in example 4.6. That is, the Kish deff says that the GREG has a variance 21.1% higher than that of a simple random sampling mean. The directly computed deff for estimated mean expenditures per hospital (**EXPTOTAL**) is 0.595. The reason for this small deff is that end-of-year patient count and beds within hospital type are powerful predictors of expenditures—a relationship that the Kish deff ignores. Consequently, the $\text{deff}_K$ is not a useful measure here. The code for this example is in `ex.4.10_smho.greg.deff.do`.

**Example 4.10: Design effect because of weighting for GREG weights.** Using the same sample as in example 4.6, we have

```
. use http://www.stata-press.com/data/svywt/smhosam, clear
. svycal regress EOYCNT i.hosptype#c.BEDS [pw = wt], gen(gregwt1)
> totals(EOYCNT=505345 1.hosptype#c.BEDS=37978 2.hosptype#c.BEDS=13066
> 3.hosptype#c.BEDS=9573 5.hosptype#c.BEDS=10077) noconstant
. svyset ID [pweight = gregwt1], strata(hosptype) regress(EOYCNT
> i.hosptype#c.BEDS, totals(EOYCNT=505345 1.hosptype#c.BEDS=37978
> 2.hosptype#c.BEDS=13066 3.hosptype#c.BEDS=9573 5.hosptype#c.BEDS=10077)
> noconstant)
.           * compute Kish deff(d) directly
. summarize gregwt1
. generate deffK = 1 + r(Var)/r(mean)^2

. display deffK
1.210932

.           * compute deff directly
. svy: mean EXPTOTAL

. estat effects, deff
```

|          |          | Linearized |          |
| -------- | -------- | ---------- | -------- |
|          | Mean     | Std. Err.  | DEFF     |
| EXPTOTAL | 1.04e+07 | 857743.1   | .594673  |

∎

Instead, we recommend $\text{deff}_K$ calculations be used as a comparative measure. For example, a survey statistician might calculate an UWE for two-sample designs to determine which one results in lower weight variability. Another example is to compare the UWEs by important analytic domains before and after a weight adjustment is applied to ensure the newly created weight is not drastically different from the input weight. If the postadjustment UWE is large relative to the input weight calculation (large being a subjective definition in the hands of the developer), then one should evaluate the contents of the adjustment model or consider tightening the bounds on the adjustment. We revisit the latter issue again in chapter 8.

The example above illustrates that having unequal weights can be very efficient in some applications. There are some alternative deffs intended to reflect reductions in variance that accrue from using estimators that have differential weights. Spencer (2000) gives a deff that accounts for any gains from pps sampling; Henry and Valliant (2015) present one that reflects gains because of calibration estimation. The Kish, Spencer, and Henry deffs are all available in the R `PracTools` package.

# 5 Use of weights in variance estimation

After sample data are collected, estimates are made and their standard errors (SEs) should be computed. The options for how to do this depend on the complexity of the sample design and estimator. The goal should be to estimate SEs that reflect sample design complexities (stratification, clustering, and method of sample selection) and complexities of the estimator (nonresponse adjustments, calibration), and type of estimator (like total, mean, or regression coefficient).

Analysts may or may not have sufficient information to compute SEs that actually reflect all of these intricacies. If you are using a file prepared by someone else, it is quite likely that the SEs you can compute will omit some sources of variability. This is often the case for public-use files provided by a government agency or academic survey organization. On the other hand, if you are preparing the data file yourself, with some work, you should be able estimate SEs that more honestly reflect the precision of estimators. We explain the options for variance estimation below, along with the circumstances where some options work better than others.

Three alternatives for computing variances and SEs for survey sample estimates are exact formulas, linearization, and replication. Särndal, Swensson, and Wretman (1992) and Wolter (2007) are good sources for mathematical details of these methods.

## 5.1 Exact formulas

Exact formulas apply only for very simple designs and estimators. For example, suppose the design is simple random sampling without replacement (srswor), and there are no further complications like nonresponse or calibration adjustments. If a mean is estimated with $\widehat{\overline{y}} = \sum_{i \in s} y_i/n$ , then its variance is estimated as

$$v\left(\widehat{\overline{y}}\right) = (1 - f)\,\widehat{S}^2/n \qquad (5.1)$$

where $\widehat{S}^2 = \sum_{i \in s} \left(y_i - \widehat{\overline{y}}\right)^2/(n-1)$ is the estimated population variance of $y$. Note that a population variance is sometimes referred to as a unit variance. The term $f = n/N$ is the sampling fraction; $1 - f$ is called the "finite population correction" factor (fpc). The final weight for each unit (which, in this case, equals the base weight) is the same, $w_i = N/n$. The estimated variance reduces as a larger fraction of the population is sampled. An estimate of a population proportion is a special case of the mean with each

$y$ equal to 1 if the unit has the characteristic and 0 if not. The estimated proportion is $\widehat{p}$, the proportion of units in the sample with the characteristic. Its estimated variance is

$$v\left(\widehat{p}\right) = (1-f)\,\widehat{p}(1-\widehat{p})/(n-1)$$

A similar formula applies if a stratified simple random sample is selected without replacement (stsrswor). For an stsrswor, we define the following terms:

- $N_h$ = the known number of units in the population in stratum $h$ $(h = 1, 2, \ldots, H$ with $H \geq 2)$;

- $W_h = N_h/N$ = the proportion of units in the population in stratum $h$;

- $n_h$ = the size of the srswor selected in stratum $h$;

- $y_{hi}$ = the value of the $y$ variable for unit $i$ in stratum $h$; and

- $s_h$ = the set of sample units in stratum $h$.

An estimated mean in stsrswor is

$$\widehat{\overline{y}}_{st} = \sum_{h=1}^{H} W_h \overline{y}_h$$

where $\overline{y}_h = \sum_{i \in s_h} y_{hi}/n_h$. When estimating a population proportion, the estimator is similar,

$$\widehat{p}_{st} = \sum_{h=1}^{H} W_h p_h$$

with $p_h$ defined the same way as $\overline{y}_h$, using a zero–one (indicator) $y$ variable. The weight for a sample unit in stratum $h$ in an stsrswor is $N_h/n_h$. The estimated variance of the stratified estimator is

$$v\left(\widehat{\overline{y}}_{st}\right) = \sum_{h=1}^{H} W_h^2 \frac{1-f_h}{n_h} \widehat{S}_h^2 \tag{5.2}$$

where $f_h = n_h/N_h$ (the stratum-specific sampling fraction) and $\widehat{S}_h^2$ is the estimated unit variance in stratum $h$.

Exact formulas can also apply for variances of simple statistics like estimated totals, means, and proportions when varying probabilities of selection are used. However, they involve joint selection probabilities, that is, the probability that two units are selected for the same sample. These joint probabilities are usually not available to analysts, either because they are not included in a file that someone else has prepared or because the sample was selected in a way that makes the joint probabilities difficult or impossible to compute. [The variances above for srswor and stsrswor actually do account for the joint selection probabilities, but the variances simplify to the nice forms in (5.1) and (5.2).]

For multistage designs, variances of simple estimators like means or totals can be written as a combination of components—one for each stage of sampling. As for without-replacement varying-probability sampling, the variance formulas for multistage designs are often extremely complicated.

## 5.2   The with-replacement workaround

Because of the complications of without-replacement variance formulas and incomplete knowledge of how a sample was actually selected, a with-replacement (WR) workaround is often used for variance estimation. This consists of using a variance estimate for with-replacement sampling, which is much simpler, even though the sample was selected without replacement. The default method for computing variances in Stata is with-replacement. To use one of the more complicated without-replacement formulas, you need to give Stata more information about the sample design. The same can be said for most software designed to analyze survey data.

The values of variance estimates from without-replacement formulas will decrease as the fraction of the population sampled becomes noticeable, that is, when the fpc is nonnegligible, say, larger than 10%. Generally speaking, this decrease is appropriate for descriptive statistics like means or totals. However, for analytic statistics like model parameter estimates using without-replacement formulas is dubious. In most situations, analysts are probably fitting a model hoping it applies beyond just the particular finite population that was sampled. In that case, a nonzero SE would be reported even if a census were done. With-replacement formulas have that property, and, as a result, are preferable for model parameter estimates and other analytic statistics.

First, consider a single-stage, stratified design in which either elements are selected directly or PSUs are selected and entirely enumerated. The probabilities of selection of the elements or PSUs are allowed to vary within strata and across strata. The `svyset` syntax for a general design is

```
svyset psu [pweight=weight], strata(strata) fpc(fpc)
```

For a one-stage design the psu variable is the field in the data file that holds the ID for a sample element; strata is the field for the stratum variable, if any; and fpc is the sampling fraction. (Note that the fpc field does give the sampling fraction, for example, $n/N$ in an srswor, and not the actual fpc, which would be $1 - n/N$ in an srswor.) If fpc is omitted, the with-replacement variance formula will be used.

The default with-replacement variance estimator used by Stata (StataCorp 2017) and several other software packages is called the "ultimate cluster" (UC) estimator (see Hansen, Hurwitz, and Madow [1953]). Suppose that $\widehat{t}_{y1}$ is an estimated total for some $y$ within a one-stage design. The UC estimator (with an ad hoc fpc added) for a one-stage, stratified sample in which $n_h$ elements or PSUs are selected from $N_h$ units in stratum $h$ is defined as

$$v(\widehat{t}_{y1}) = \sum_{h=1}^{H}(1-f_h)\frac{n_h}{n_h-1}\sum_{i=1}^{n_h}\left(w_{hi}y_{hi} - \frac{1}{n_h}\sum_{j=1}^{n_h}w_{hj}y_{hj}\right)^2 \qquad (5.3)$$

where $f_h = n_h/N_h$ is the sampling fraction in stratum $h$. In the special case of srswor, (5.3) reduces to exactly (5.1). In stsrswor, (5.3) equals (5.2). Without the fpc, (5.3) equals a variance that is appropriate when the sample is selected with varying probabilities and with-replacement if $w_{hi}$ is the inverse of the selection probability for unit $hi$. Base weights (and not, say, base weights adjusted for nonresponse) are required to satisfy the definition of a linear point estimator $\widehat{t}_{y1}$; we discuss nonlinear estimators in section 5.3.

The example below uses a stratified simple random sample selected without replacement from the Survey of Mental Health Organizations population (smho.n874.dta) in the R package PracTools (Valliant, Dever, and Kreuter 2016). Thirty hospitals were selected from each of four strata defined by hospital type (hosptype). The weight field is wt, and the stratum population count is in Nh. Totals are estimated for number of patients seen in a calendar year (SEENCNT) and number of inpatient beds (BEDS). In the first table below, SEs are estimated using (5.2) without an fpc. In the second, a stratum-specific fpc is included. As the comparison shows, SEs are somewhat smaller when the fpc is incorporated in the variance estimates.

**Example 5.1: Variance estimation in an stsrswor sample.** The code for this example is in ex.5.1_stsrswor.do.

```
. use http://www.stata-press.com/data/svywt/smhosam

.           * No fpc
. svyset ID [pw=wt], strata(hosptype)

. svy: total SEENCNT BEDS, cformat(%9.0gc)
(running total on estimation sample)
Survey: Total estimation

Number of strata =        4         Number of obs   =        120
Number of PSUs   =      120         Population size =        725
                                    Design df       =        116
```

|          | Total    | Linearized Std. Err. | [95% Conf. Interval] |          |
|---------:|---------:|---------------------:|---------------------:|---------:|
| SEENCNT  | 1210148  | 139,295              | 934,257              | 1486040  |
| BEDS     | 58,671.6 | 5,876.78             | 47,031.9             | 70,311.3 |

```
        .                 * with fpc
        . svyset ID [pw=wt], strata(hosptype) fpc(Nh)


        . svy: total SEENCNT BEDS, cformat(%9.0gc)
        (running total on estimation sample)

        Survey: Total estimation

        Number of strata =        4        Number of obs   =         120
        Number of PSUs    =      120        Population size =         725
                                            Design df       =         116
```

|         |          | Linearized |         |              |
|---------|---------:|-----------:|--------:|-------------:|
|         |    Total | Std. Err.  | [95% Conf. | Interval] |
| SEENCNT | 1210148  | 124,058    | 964,435 | 1455862      |
| BEDS    | 58,671.6 | 5,152.93   | 48,465.5 | 68,877.6    |

■

For a multistage, stratified sample where $m_h$ out of $M_h$ PSUs are selected in stratum $h$ and $n_{hi}$ elements are selected within sample PSU $hi$, the estimated total is $\widehat{t}_{ym} = \sum_{h=1}^{H} \sum_{i=1}^{m_h} \sum_{j=1}^{n_{hi}} w_{hij} y_{hij}$. The term $w_{hij}$ is the unit-level analysis weight calculated as a function of the base weight $(\pi_i \pi_{j|i})^{-1}$ (see section 2.1) and possibly one or more adjustments for coverage and nonresponse for one or multiple stages of the design. The UC variance estimator is

$$v(\widehat{t}_{ym}) = \sum_{h=1}^{H} (1 - f_h) \frac{m_h}{m_h - 1} \sum_{i=1}^{m_h} \left( y_{hi}^* - \widehat{\overline{y}}_h^* \right)^2 \tag{5.4}$$

where $f_h = m_h / M_h$, $y_{hi}^* = \sum_{j=1}^{n_{hi}} w_{hij} y_{hij}$, and $\widehat{\overline{y}}_h^* = \sum_{i=1}^{m_h} y_{hi}^* / m_h$. The estimator in (5.3) is a special case of (5.4) when there are no clusters in any stratum.

Removing the fpc from (5.4), we have a variance estimator that is appropriate for a design in which PSUs are selected with replacement in each stratum. If PSUs are selected WR, (5.4) without the fpc is appropriate if the within-PSU sampling permits an unbiased estimator of the total for each PSU to be constructed. Thus, (5.4) is applicable to many types of within-PSU sample design, for example, srswor, stsrswor, and sampling with varying probabilities without replacement. In many public-use files, an analyst's only choice will be to use WR variance estimates because of a lack of information to do otherwise. Note that this is acceptable to many analysts because the resulting SEs may be larger than necessary (that is, conservative) but hopefully only slightly so.

**Example 5.2: Stratified multistage public-use file.** The following example gives the code for estimating the total number of persons in the United States by race and ethnicity (RIDRETH1) from the 2009–2010 National Health and Nutrition Examination Survey (NHANES). The method of variance estimation is specified with no fpc as vce(linearized). (Linearization variances are described in the next section.) This leads to (5.4) being used with $f_h = 0$, that is, the WR estimate. The code is in ex.5.2_linvar.nhanes.do.

```
. use http://www.stata-press.com/data/svywt/nhanes2009.10.dta

. svyset SDMVPSU [pw=WTMEC2YR], strata(SDMVSTRA) vce(linearized)

. svy: tabulate RIDRETH1, format(%14.2fc) count se cv stubwidth(25)
(running tabulate on estimation sample)
Number of strata    =         15        Number of obs    =        5,001
Number of PSUs      =         31        Population size   = 192,811,171
                                        Design df         =           16


                  RIDRETH1 |     count               se                cv
---------------------------+-----------------------------------------------
          Mexican American | 17,661,953.97     3,681,216.94           20.84
            Other Hispanic | 10,352,341.27     2,425,211.03           23.43
        Non-Hispanic White | 127,649,401.95   12,272,064.45            9.61
        Non-Hispanic Black | 22,765,771.83     2,038,826.29            8.96
   Other-Incl. Multi-Racial | 14,381,701.71     2,283,505.57           15.88

                     Total | 192,811,170.73
---------------------------+-----------------------------------------------
  Key:  count     =  weighted count
        se        =  linearized standard error of weighted count
        cv        =  coefficients of variation of weighted count
```

■

The degrees of freedom are 16 because the public-use file has 14 VarStrat (variance strata identical to or defined as a combination of the design strata) with 2 VarUnits (variance PSUs equal to the original or a combined version of the design PSUs) in each and 1 VarStrat with 3 VarUnits. (VarStrat and VarUnits are discussed further in section 5.4.) Note that NHANES is actually poststratified by race; if that were accounted for, the SEs above would be 0.

Expressions (5.3) and (5.4) make it clear how the weights enter the variance estimate. The fpcs in these formulas are ad hoc additions for most sample designs. Strictly speaking, fpcs are only appropriate to a limited number of sample designs. Examples are srswor, stsrswor, and two-stage cluster samples in which each stage is selected via srswor. In probability sampling designs where units are selected with varying probabilities and without replacement, using an fpc is not mathematically correct but may be a rough way of accounting for gains in precision because of its having a large sampling fraction. Stata (like other software packages) offers the option of incorporating an fpc even if variance formula for your design does not need one.

## 5.3  Linearization variances

Expression (5.4) is appropriate for a linear estimator of a total, that is, a weighted sum of $y$'s and the weights do not involve any adjustments like the weights for nonresponse. Many estimators using survey data are nonlinear. Even estimating a mean as $\widehat{\bar{y}} = \widehat{t}_y / \widehat{M}$ with $\widehat{M} = \sum_{h=1}^{H} \sum_{i=1}^{m_h} \sum_{j=1}^{m_{hi}} w_{hij}$ creates a nonlinear estimator, which requires specialized variance estimation methods. Nonresponse adjustments also introduce nonlinearity into the estimators through $1/\widehat{\phi}$ (see chapter 3).

The linearization, or Taylor series (approximation) method, is one approach to estimating the variance of a nonlinear estimator. The general idea is to approximate the nonlinear statistic with a first-order Taylor series and then compute the variance of that approximation. For example, the ratio mean $\widehat{\overline{y}}$ is a function of two estimated totals, $\widehat{t}_y$ and $\widehat{M}$. The linear approximation used to create a variance estimator is

$$\widehat{\overline{y}} - \mu = \sum_{h=1}^{H} \sum_{i=1}^{m_h} \sum_{j=1}^{n_{hi}} w_{hij} z_{hij} \equiv \widehat{z}$$

where $z_{hij} = M^{-1}(y_{hij} - \mu)$, $\mu$ is the population mean, and $M$ is the population count of elements. Assuming that neither $\mu$ nor $M$ are known, estimates are used in the formula for $z_{hij}$. The term $z_{hij}$ is variously known as a "linear substitute", "linearized variable", or "score variable". The approximate variance of the ratio mean is the variance of the estimated total, $\widehat{z}$.

Estimators of the parameters in models like the logistic and probit binary regression models must be found iteratively. The estimator of the covariance matrix produced by Stata is a type of linearization estimator called a sandwich estimator (see Binder [1983]).

Following up the NHANES example in the previous section, we estimate the proportions of persons in each race group. The estimated proportions are nonlinear because the estimated total of persons in the population (the sum of all weights) is in the denominator of each proportion, that is, $\widehat{p}_g = \sum_{i \in s_{\mathrm{ER}}} \delta_g w_i / \sum_{i \in s_{\mathrm{ER}}} w_i$ for input weights $w_i$ and $\delta_g = 1$ if respondent $i$ is in the $g$th race group (0 otherwise). Because the count option is omitted from svy: tabulate, estimated proportions are produced by default.

```
. svy: tabulate RIDRETH1, cell se cv stubwidth(25)
(output omitted)
```

| RIDRETH1 | proportion | se | cv |
|---|---|---|---|
| Mexican American | .0916 | .0226 | 24.65 |
| Other Hispanic | .0537 | .0132 | 24.56 |
| Non-Hispanic White | .662 | .0348 | 5.262 |
| Non-Hispanic Black | .1181 | .009 | 7.646 |
| Other-Incl. Multi-Racial | .0746 | .012 | 16.08 |
| Total | 1 | | |

```
Key:   proportion  =  cell proportion
       se          =  linearized standard error of cell proportion
       cv          =  coefficients of variation of cell proportion
```

## 5.4   Replication variances

The general approach in replication variance estimation is to divide the full sample into subsamples (or replicates), repeat all weighting steps separately for each replicate, compute the estimate for each replicate, and then plug those estimates into a fairly simple variance formula. The three alternatives we cover here are the jackknife, balanced repeated replication (BRR), and bootstrap. The finite population theory for the jackknife is given in Krewski and Rao (1981) and Rao and Wu (1985). BRR was introduced by McCarthy (1969) with more advanced theory presented in Krewski and Rao (1981) and Rao and Wu (1985). The bootstrap for finite populations is covered in Rao and Wu (1988).

Properly implemented, replication variances have good design-based and model-based properties. The large sample design-based theory is mainly derived for with-replacement sampling of first-stage units. Replication variances have good model-based properties under the model for which a point estimator (for example, mean or total) is exactly or approximately model unbiased. If multiple weighting steps are used (for example, nonresponse adjustments and calibration) and those steps are repeated separately for each replicate, replication variance estimators will implicitly account for the combined effects of the steps. In fact, not repeating the steps for each replicate leads to biased variance estimators. Valliant (2004), for example, demonstrates that generating replicates with only the respondent data and final weights can severely underestimate SEs by ignoring the variability of the weight adjustments such as nonresponse. Section 5.5 gives an example of reflecting the effects of both nonresponse adjustment and calibration.

Several public-use datasets produced by government agencies come with replicate weights. The American Community Survey (ACS) is released with a specialized version called successive difference replicate weights.[1] The American Housing Survey uses Fay's method of BRR in its public-use files, as does the Current Population Survey's Annual Social and Economic Supplement (ASEC).[2] The California Health Interview Survey supplies jackknife weights with its files.[3] Supplying replicate weights allows agencies to provide a method of variance estimation that works for most statistics and accounts for the complexities of sample designs and estimators. Replication weighting also does not require that users be given strata and PSU identifications, which helps protect the anonymity of sample members.

How the replicates are formed varies depending on the method. There are pros and cons for the methods, which are sketched in the upcoming sections along with examples of how to use the techniques in Stata. The undocumented procedure `svygen` will create weights for several of the replication methods. Another choice is the package `svr` (Winter 2002), which has more features than `svygen`. The files for `svr` can be installed by typing `ssc install svr` at the command line. Ancillary Hadamard matrix files can be installed by typing `net get survwgt` at the command line.

---

1. See https://usa.ipums.org/usa/repwt.shtml.
2. See http://thedataweb.rm.census.gov/ftp/cps_ftp.html#cpsrepwgt.
3. See http://healthpolicy.ucla.edu/chis/about/Pages/about.aspx.

## 5.4.1  Jackknife

The jackknife gives a consistent variance estimator (that is, one that converges to the correct value in large samples) for linear and most nonlinear estimators. Exceptions are quantiles, like the median, where the jackknife is not consistent and should not be used. In the basic delete-one jackknife, replicates are formed by dropping one PSU at a time. Dropping a PSU means that all sample elements in the PSU are dropped. For example, if the sample consists of schools and students within schools, then one school and all the sampled students within it are dropped to form a replicate. It would be a mistake to drop one student at a time.

If the sample has $m = \sum_{h=1}^{H} m_h$ PSUs, then $m$ replicates are created. When PSU $hi$ is dropped and $d_j$ is the base weight for some element $j$, the adjusted base weight for that element is

$$d_{j(hi)} = \begin{cases} 0 & \text{if unit } j \text{ is in PSU } i \text{ in stratum } h \\ \frac{m_h}{m_h-1}d_j & \text{if unit } j \text{ is in stratum } h \text{ but not in PSU } i \\ d_j & \text{if unit } j \text{ is not in stratum } h \end{cases} \qquad (5.5)$$

In other words, all units in the deleted PSU $hi$ have their weights set to 0. All units in the other $(m_h - 1)$ PSUs within stratum $h$ have their base weights multiplied by $m_h/(m_h - 1)$. Units in the strata where no PSU was dropped retain their original weight. Thus, the weights for the retained units in stratum $h$ are adjusted to represent the full stratum and the weights for units in other strata are left alone. Note that (5.5) is applicable for single-stage designs where the PSU is a single element and not just for PSUs defined as clusters of elements.

All $m$ PSUs are cycled through to create a series of $m$ jackknife base weights for each sample element. Ideally, any other steps like adjustment for unknown eligibility, adjustment for nonresponse, and calibration are then done separately for each set of replicates. These adjustments are applied in Stata and other software by "looping" over the variance replicates and applying the same program code used for the full-sample adjustments. There are no special functions to perform the calculations outright. A series of $m+1$ final weights is then appended to each record in the data file—one for the full sample and one for each jackknife replicate. The jackknife weights are supplied to svyset with the jkrweight() option as illustrated below. Stata can also compute the adjustments in (5.5) and apply them to the full-sample weight using svygen. However, be aware that this adjustment is too crude to reflect the effects on variances of multiple weighting steps.

The variance formula for the stratified delete-one jackknife is

$$v_J\left(\widehat{\theta}\right) = \sum_{h=1}^{H}\left(1 - f_h\right)\frac{m_h - 1}{m_h}\sum_{i=1}^{M_h}\left(\widehat{\theta}_{(hi)} - \widehat{\theta}_h^*\right)^2 \tag{5.6}$$

where $\widehat{\theta}_{(hi)}$ is the replicate estimate when PSU $hi$ is dropped out. The term $f_h$ is the sampling fraction of first-stage units in stratum $h$ and, as in (5.3), is an ad-hoc addition except in stsrswor designs where it follows from theory. This estimator is sometimes called the JKn jackknife. Two choices are available in Stata for the term $\theta_h^*$. When the mse option is specified in svyset, $\widehat{\theta}_h^*$ is the full-sample estimate. Otherwise, $\widehat{\theta}_h^*$ is the mean of the delete-one estimates in stratum $h$. The mse choice will generally lead to larger estimated SEs, although, in theory, both estimate the true variance. A special case of (5.6) is the unstratified delete-one jackknife, which is sometimes called JK1.

Using the smhosam.dta stsrswor sample, the weights in (5.5) can be created and used to estimate the totals and means per hospital of patients seen and number of beds with the following code, which is in ex.5.3_jKn.smhosam.do. The jackknife weights are labeled jkw1–jkw120. The fpc(Nh) option of svygen includes an fpc in the variance estimates.

**Example 5.3: JKn example.**

```
. use http://www.stata-press.com/data/svywt/smhosam, clear
. svygen jackknife jkw [pw = wt], strata(hosptype) psu(ID) fpc(Nh)
.    * check some JKn weights
. tabulate hosptype, summarize(jkw1)
. tabulate hosptype, summarize(jkw100)
. svyset ID [pw=wt], jkrweight(jkw*) strata(hosptype) vce(jackknife) mse
. svy: total SEENCNT BEDS, cformat(%9.0gc)

. svy: mean SEENCNT BEDS, cformat(%9.0gc)
(running mean on estimation sample)
Jackknife replications (120)
————+——— 1 ———+——— 2 ———+——— 3 ———+——— 4 ———+——— 5
.................................................      50
.................................................     100
..................
Survey: Mean estimation

Number of strata =          4        Number of obs      =          120
                                     Population size    =          725
                                     Replications       =          120
                                     Design df          =          116
```

|          |   Mean   | Jknife *<br>Std. Err. | [95% Conf. Interval] |          |
|----------|----------|-----------------------|----------------------|----------|
| SEENCNT  | 1,669.17 | 171.1151              | 1,330.26             | 2,008.09 |
| BEDS     | 80.9263  | 7.107493              | 66.84901             | 95.00358 |

As the calculation occurs, a dot will be printed as the program cycles through each replicate. In the presence of many replicates, the dots indicate that something is happening and the program has not stopped working.

If we had not used an fpc, the standard errors for the estimated totals (not shown above) would be exactly the same as the with-replacement standard errors in section 5.2 where the default linearization variances were illustrated. This equality is as it should be, because the delete-one jackknife variance estimator is exactly the same as the stsrswr variance when the estimator is linear. The jackknife and linearization SEs for the means are also quite close for this example—a fact that the reader can check.

**Grouped jackknife**. In some surveys, the number of sample PSUs and the number of replicates in the delete-one jackknife is very large. Single-stage samples like telephone surveys of households or surveys of establishments can have thousands of sample units. For example, a single-stage survey of 1,000 households would require 1,000 replicates for the full delete-one jackknife. In such a case, a group of PSUs may be dropped to form a replicate. To reduce the number of replicates, PSUs within strata or strata themselves are often combined. We refer to a combined stratum as a VarStrat, and we refer to a combined PSU as a VarUnit. There are many references that explore the properties of the grouped jackknife variance estimator, for example, Kott (1999); Kott (2001); Lu, Brick, and Sitter (2006) and Wolter (2007, chap. 4).

Section 15.5 in Valliant, Dever, and Kreuter (2013) summarizes some of the considerations in grouping PSUs (which we will not cover here). The general idea is to form groups so that each replicate estimate is a legitimate estimate of the full population quantity. If a group of PSUs is deleted, the weight adjustment for a retained element in VarStrat $h$ and VarUnit $g$ should be $m_h / (m_h - m_{hg})$, where $m_h$ is the number of PSUs in VarStrat $h$ and $m_{hg}$ is the number of PSUs in VarUnit $hg$ that are dropped (see also Valliant, Brick, and Dever [2008]). In other words, the weight adjustment depends on how many original PSUs are dropped. If every group has the same number of PSUs, then this weight adjustment reduces to $G_h / (G_h - 1)$, where $G_h$ is the number of groups in VarStrat $h$. That is, the adjustment becomes a particular case of (5.5). If the VarStrat are the original design strata, then the weight adjustment is $m_h / (m_h - 1)$, as in (5.5). Stata will not compute these fine-tuned adjustments for the grouped jackknife—the adjusted weight variables must be supplied to `svyset` with the `jkrweight()` option.

The jackknife variance formula used by Stata that applies to the delete-one-group case is

$$v_J\left(\widehat{\theta}\right) = \sum_{h=1}^{H} (1 - f_h) K_h \sum_{g=1}^{G_h} \left(\widehat{\theta}_{(hg)} - \widehat{\theta}_h^*\right)^2$$

where $H$ is the number of VarStrat, $K_h$ is a multiplier that depends on whether the jackknife is grouped, $\widehat{\theta}_{(hg)}$ is the estimate computed after deleting VarStrat or VarUnit $hg$. If no combining of design strata is done, then $h$ just denotes an original design stratum. For the delete-one-PSU jackknife, $K_h = (m_h - 1)/m_h$ and is equivalent to (5.6). For the delete-a-group jackknife, $K_h = (m_h - m_{hg})/m_h$. The field that holds $K_h$ is specified with the `multiplier()` suboption of `jkrweight()`.

To illustrate the use of the unstratified jk1 with grouping of PSUs, we use the same dataset as in examples 4.3, 4.4, and 4.5. The field JKgrp holds the group identifiers used to create 50 jackknife subsamples.

**Example 5.4: Raking with survwgt.** The survwgt command by Winter (2002) will compute raked weights for each jk1 replicate. (See ex.5.4_rake.survwgt.do for the code.) After recoding hisp into hispr as in the previous examples, the replication weights shown in (5.5) are calculated with the command

```
survwgt create jk1, psu(JKgrp) weight(wt) stem(JK1_)
```

The weights for each replicate are stored with the prefix JK1_. The marginal totals for age groups and recoded Hispanic are stored in age_tot and hispr.

```
. use http://www.stata-press.com/data/svywt/nhis_sam, clear
. label define age_lab 1 "<18" 2 "18-24" 3 "25-44" 4 "45-64" 5 "65+"
. label values age_grp age_lab
. recode hisp (1=1) (2=2) (3=3) (4=3), gen(hispr)
.         * create JK1 replicate weights
. survwgt create jk1, psu(JKgrp) weight(wt) stem(JK1_)
. generate age_tot = 5991 if age_grp == 1
. replace age_tot = 2014 if age_grp == 2
. replace age_tot = 6124 if age_grp == 3
. replace age_tot = 5011 if age_grp == 4
. replace age_tot = 2448 if age_grp == 5
. generate hispr_tot =  5031 if hispr == 1
. replace hispr_tot = 12637 if hispr == 2
. replace hispr_tot =  3920 if hispr == 3
. survwgt rake [all], by(age_grp hispr) totvars(age_tot hispr_tot) stem(JKrake)
```

```
. svrtab age_grp hispr, count se format(%12.0f)

Cross-tabulation with replication-based (jk1) standard errors

Analysis weight:     JKrake0          Number of obs       =       2000
Replicate weights:   JKrake1...       Population size     =      21588
Number of replicates: 50              Degrees of freedom  =         49
```

|          | RECODE of hisp (hisp) |        |        |        |
|---------:|---------:|---------:|---------:|---------:|
| age.grp  |    1     |    2     |    3     | Total  |
|      <18 |  1869    |  2878    |  1243    |  5991  |
|          |   (88)   |  (120)   |   (98)   |   (0)  |
|    18-24 |   612    |  1041    |   361    |  2014  |
|          |   (92)   |  (101)   |   (72)   |   (0)  |
|    25-44 |  1600    |  3448    |  1076    |  6124  |
|          |  (104)   |  (107)   |   (77)   |   (0)  |
|    45-64 |   716    |  3350    |   945    |  5011  |
|          |   (68)   |   (92)   |   (70)   |   (0)  |
|      65+ |   234    |  1919    |   295    |  2448  |
|          |   (51)   |   (69)   |   (50)   |   (0)  |
|    Total |  5031    | 12637    |  3920    | 21588  |

```
  Key:  weighted counts
        (standard errors of weighted counts)

  Pearson:
    Uncorrected    chi2(8)        =     95.5011
    Design-based   F(6.28, 307.67) =      9.6460       P = 0.0000
```

The key command that accomplishes the raking separately for each replicate and stores the raked jackknife weights with the stem JKrake is

```
survwgt rake [all], by(age_grp hispr) totvars(age_tot hispr_tot) stem(JKrake)
```

The variable specification [all] indicates that the raking should be done for the full sample and for every replicate. Behind the scenes, survwgt has issued its own updated equivalent to svyset. As the above table generated by svrtab shows, the SEs for the estimated marginal totals of age_grp and hispr are all 0, as they should be, because there will be no sample-to-sample variation in the estimates. Note that the calculations were conducted without specifying the original design strata or PSUs.

■

Next, we use the California Health Interview Survey (CHIS) to illustrate another variation on the jackknife. CHIS is a telephone survey of California residents that has been conducted every year since 2001. CHIS collects extensive information for all age groups on health status, health conditions, health-related behaviors, health insurance coverage, access to health care services, and other health and health-related issues. Once a house-

hold was sampled, an adult within that household was sampled. If there were children
or adolescents in the household, one child or one adolescent was eligible for sampling.
CHIS has several sample design and estimation complexities, including different sampling
rates by county. This way, one can make reliable county and state estimates. Other
complexities are within household sampling of one adult, household- and person-level
nonresponse adjustments, and raking to population control totals by geographic area,
age, gender, race, and various crosses of these variables (Flores-Cervantes and Brick
2014; Flores-Cervantes, Norman, and Brick 2014).

Because of these complexities, a form of jackknife replication called JK2 was used
(see Flores-Cervantes et al. [2014, sec. 9.3]). JK2 assumes that two PSUs are selected
per stratum. Replicates are formed by randomly dropping one PSU in each stratum.
The method is described in more detail in Westat (2007, app. A). In JK2, $K_h = 1$ and
$\widehat{\theta}_h^* = \widehat{\theta}$. Each estimation step, including the nonresponse adjustment and raking to
population controls, was repeated separately for each replicate.

In the CHIS adults 2011–2012 dataset that accompanies this book, the full-sample
raked weight is called `rakedw0`. There are 80 replicate raked weights denoted by
`rakedw1` - `rakedw80`. In the `svyset` command, `multiplier(1)` is used in the option `jkrweight()` to set $K_h = 1$. We also use the option `dof(80)` to set the degrees of
freedom (df) to 80. This is appropriate because there are 80 VarStrat and 2 VarUnits
in each. (We pick up 1 df from each VarStrat.) If `dof(80)` was not used, the design df will be computed as 79. In this case, the practical effect is minimal. (See
`ex.5.5_jk2_chis.do` for the Stata code.)

### Example 5.5: JK2 example with raked weights.

```
. use http://www.stata-press.com/data/svywt/adult, clear
. svyset [pw = rakedw0], jkrweight(rakedw1 - rakedw80, multiplier(1)) vce(jack)
> dof(80) mse
      pweight: rakedw0
          VCE: jackknife
          MSE: on
    jkrweight: rakedw1 .. rakedw80
  (output omitted )
```

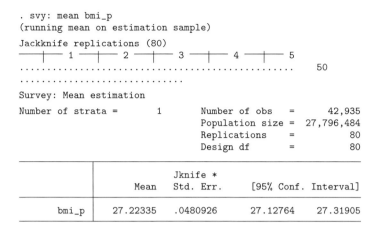

```
. svy: mean bmi_p
(running mean on estimation sample)
Jackknife replications (80)
```

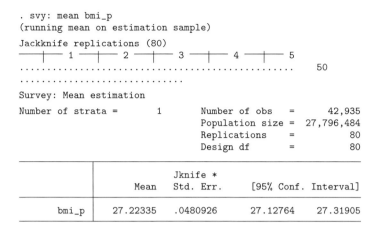

```
Survey: Mean estimation

Number of strata =        1      Number of obs    =      42,935
                                 Population size = 27,796,484
                                 Replications     =          80
                                 Design df        =          80
```

|  |  | Jknife * |  |  |
|---|---|---|---|---|
|  | Mean | Std. Err. | [95% Conf. Interval] | |
| bmi_p | 27.22335 | .0480926 | 27.12764 | 27.31905 |

The mse option in svyset centers the replicate estimates around the full-sample estimate in the variance calculation. Otherwise, the average of the replicate estimates is used.

■

## 5.4.2   Balanced repeated replication

Balanced repeated replication (BRR), or balanced half-sampling, was developed by Mc-Carthy (1969) for sample designs where two PSUs are selected from each stratum. BRR gives consistent variance estimates for linear and nonlinear estimates. Unlike the jack-knife, BRR can be used for percentiles such as the median and quartiles.

This type of design is often used in area samples where strata are geographic regions within a country. Sampling only two PSUs from each stratum ensures that the sample will be spread across a country. You can take even more control over the distribution of the sample by selecting one PSU from each stratum. In a 1-PSU per stratum design, strata are usually combined for variance estimation; thus, a 2-PSU per stratum setup to which BRR can be applied is created. Additionally, this approach is not limited to area sampling. The CHIS survey is an example of a telephone survey where sample households are grouped together into a 2-VarUnit per VarStrat design. BRR can then be applied to that case also.

**Standard BRR**. Each replicate subsample consists of one PSU selected from each stratum in a prescribed way. A Hadamard matrix is used to identify a set of so-called "balanced" half-samples, which is much smaller than $2^H$, the number of possible randomly selected half-samples for a design with $2 \times H$ PSUs. (Ancillary Hadamard matrix files are installed using net get survwgt at the command line.) The number of half-samples in a balanced set is the smallest multiple of four that is greater than or equal to the number of strata.

Hadamard matrices are usually represented by +1s and −1s. A $4 \times 4$ example is

$$H = \begin{pmatrix} +1 & +1 & +1 & +1 \\ +1 & -1 & +1 & -1 \\ +1 & +1 & -1 & -1 \\ +1 & -1 & -1 & +1 \end{pmatrix}$$

Rows are for strata; columns are for replicates. The first column having all +1s means that the first PSU from each stratum should be selected for replicate 1. The second column of $(+1, -1, +1, -1)$ means that the second replicate contains PSU 1 from stratum 1, PSU 2 from stratum 2, PSU 1 from stratum 3, and PSU 2 from stratum 4. If $H = 4$, the number of replicates needed for an orthogonal set is 4.

In the standard BRR, weights for the units retained in a replicate are multiplied by 2; others are set to 0. Consequently, the adjusted weights for the units in replicate $\alpha$ are

$$d_{0k(\alpha)} = \begin{cases} 0 & \text{if unit } k \text{ is in a PSU that is not in the half sample} \\ 2d_{0k} & \text{if unit } k \text{ is in a PSU that is in the half sample} \end{cases}$$

The adjusted weights are then used to compute a replicate estimate denoted by $\widehat{\theta}_\alpha$. With $\alpha = 1, \ldots, A$ replicates, the BRR variance estimator is

$$v_{\text{BRR}}\left(\widehat{\theta}\right) = A^{-1} \sum_{\alpha=1}^{A} \left(\widehat{\theta}_\alpha - \widehat{\theta}^*\right)^2 \tag{5.7}$$

If the option `mse` is used in `svyset`, then $\widehat{\theta}^*$ is the full-sample estimate. If `mse` is omitted, then $\widehat{\theta}^*$ is the average of the half-sample estimates.

More complicated estimators, which can be functions of several estimated totals, are treated similarly. If the full-sample estimate is a function of $p$ estimated totals, that is, $\widehat{\theta} = f\left(\widehat{t}_1, \ldots, \widehat{t}_p\right)$, then a replicate estimate is $\widehat{\theta}_\alpha = f\left(\widehat{t}_{1\alpha}, \ldots, \widehat{t}_{p\alpha}\right)$ where $\widehat{t}_{j\alpha}$ is the estimated total for the $j$th variable based on the units in half-sample $\alpha$.

**Fay–BRR.** A drawback of the standard BRR is that units in domains with small samples may be absent or nearly so in some replicates, which is a condition that will lower the precision of the domain variance estimates. A solution to this problem is the Fay–BRR method (Fay 1984; Judkins 1990), in which units are down-weighted rather than completely dropped. Half-samples are identified using a Hadamard matrix as above. The weights are then calculated as

$$d_{0k(\alpha)} = \begin{cases} \rho d_{0k} & \text{if unit } k \text{ is in a PSU that is not in the half sample} \\ (2 - \rho)\, d_{0k} & \text{if unit } k \text{ is in a PSU that is in the half sample} \end{cases}$$

where $0 \le \rho < 1$. If $\rho = 0$, this is the standard BRR. The form of the Fay–BRR variance estimator is

$$v_{\text{F-BRR}}\left(\widehat{\theta}\right) = \frac{1}{A(1-\rho)^2} \sum_{\alpha=1}^{A} \left(\widehat{\theta}_\alpha - \widehat{\theta}^*\right)^2 \tag{5.8}$$

Judkins (1990) recommends perturbation factors ($\rho$) in the range of 0.5 and 0.7 for a general-purpose analysis file. These values produced stable results with low bias within the criteria included in his simulation study.

The next example creates standard BRR and Fay–BRR weights for the 2009–2010 NHANES dataset first described in example 5.1. This dataset has 15 VarStrat (analytic strata), all but one of which has two sample PSUs (VarUnits). In the stratum with three PSUs, the PSU variable (SDMVPSU) is recoded to have only two values for the sake of this example. The matrix code below creates a $16 \times 16$ Hadamard matrix using the fact that if $H$ is a Hadamard matrix, then so is

$$\begin{pmatrix} H & H \\ H & -H \end{pmatrix}$$

The matrix code below uses the Kronecker product, symbolized by #, to construct the matrix. `svygen brr` creates the replicate weights. When the `fay(0.5)` option is used, Fay–BRR weights are created with $\rho = 0.5$. The mean difference between inspiration and exhalation chest circumstance (ARDDINEX) is estimated using `svy: mean`. Chest expansion measurement is a clinical measure to assess limitation of thoracic spine mobility. Limited mobility is an effect of spinal arthritis or spondyloarthritis. In this example, the values of the standard and Fay–BRR SEs are very similar. Note that the Fay–BRR $\rho$ value must be specified both when generating the weights in `svygen` and in `svyset` where the weights to use for estimation are specified. For both the standard and Fay–BRR, the `mse` option in `svyset` causes the variances in (5.7) and (5.8) to be centered around the full-sample estimate. The code is in `ex.5.6_brr.nhanes.do`.

## Example 5.6: BRR example using svygen.

```
. use http://www.stata-press.com/data/svywt/nhanes2009.10.dta
.        * combine SDMVPSU = 2 and 3 in one stratum to use BRR
. generate newVPSU = SDMVPSU
. recode newVPSU (3=2)
.        * create Hadamard matrix of dimension 16
. matrix h2 = (-1, 1 \ 1, 1)
. matrix h4 = h2 # h2
. matrix h16 = h4 # h4
. svygen brr stdBRRwts [pw=WTMEC2YR], strata(SDMVSTRA) psu(newVPSU)
> hadamard(h16)
. svygen brr fayBRRwts [pw=WTMEC2YR], strata(SDMVSTRA) psu(newVPSU)
> hadamard(h16) fay(0.5)
.     * Mean difference in inspiration/exhalation chest circumference
. svyset newVPSU [pw=WTMEC2YR], brrweight(stdBRR*) vce(brr) mse
```

```
. svy: mean ARDDINEX
(running mean on estimation sample)
BRR replications (16)
──────┼──── 1 ──────┼──── 2 ──────┼──── 3 ──────┼──── 4 ──────┼──── 5
. . . . . . . . . . . . . . .
Survey: Mean estimation          Number of obs    =         4,615
                                 Population size = 179,833,547
                                 Replications     =            16
                                 Design df        =            15
```

|          |          | BRR *     |                      |          |
|---------:|---------:|----------:|---------------------:|---------:|
|          | Mean     | Std. Err. | [95% Conf. Interval] |          |
| ARDDINEX | 4.777534 | .0539973  | 4.662441             | 4.892626 |

```
. svyset newVPSU [pw=WTMEC2YR], brrweight(fayBRR*) vce(brr) fay(0.5) mse

. svy: mean ARDDINEX
  (output omitted)
```

|          |          | BRR *     |                      |          |
|---------:|---------:|----------:|---------------------:|---------:|
|          | Mean     | Std. Err. | [95% Conf. Interval] |          |
| ARDDINEX | 4.777534 | .0529426  | 4.664689             | 4.890378 |

■

An alternative to svygen is the program survwgt (Winter 2002), which creates several kinds of replication weights, including the standard BRR and Fay–BRR. The example below repeats the NHANES calculation above with the same recoding of the PSU variable. The survwgt program has its own Hadamard matrix files that, by default, are stored as brr_hadamardmatrixfile.ado. When BRR weights are created, survwgt locates the Hadamard matrix of appropriate size given the number of strata. Note that svrset and svrmean are used instead of svyset and svy: mean. See ex.5.7_brr.survwgt.do for the full code.

### Example 5.7: BRR example using survwgt.

```
. survwgt create brr, strata(SDMVSTRA) psu(newVPSU) weight(WTMEC2YR)
> stem(stdBRR)

. svrset set meth brr pw WTMEC2YR rw stdBRR1-stdBRR16

. svrmean ARDDINEX
Survey mean estimation, replication (brr) variance method

Analysis weight:     WTMEC2YR             Number of obs      =       4615
Replicate weights:   stdBRR1...           Population size    = 1.798e+08
Number of replicates: 16                  Degrees of freedom =         15
k (Fay's method):    0.000
```

| Mean | Estimate | Std. Err. | [95% Conf. Interval] | | Deff |
|------|----------|-----------|--------|--------|------|
| ARDDINEX | 4.777534 | .0528751 | 4.664833 | 4.890234 | 3.237466 |

```
. survwgt create brr, strata(SDMVSTRA) psu(newVPSU) weight(WTMEC2YR) fay(0.5)
> stem(fayBRR)

. svrset set meth brr pw WTMEC2YR rw fayBRR1-fayBRR16 fay 0.5

. svrmean ARDDINEX
Survey mean estimation, replication (brr) variance method

Analysis weight:     WTMEC2YR             Number of obs      =       4615
Replicate weights:   fayBRR1...           Population size    = 1.798e+08
Number of replicates: 16                  Degrees of freedom =         15
k (Fay's method):    0.500
```

| Mean | Estimate | Std. Err. | [95% Conf. Interval] | | Deff |
|------|----------|-----------|--------|--------|------|
| ARDDINEX | 4.777534 | .0523896 | 4.665868 | 4.889199 | 3.178291 |

■

By default, survwgt will print a dot for each replicate as it creates weights. This assures that something is happening, but you can suppress the dots with the nodots option.

As for the jackknife, practitioners may combine strata or PSUs to form VarStrat and VarUnits to limit the number of replicates and hence the size of the analytic data files. In that case, BRR can be applied to the combined units. An example is the 2009 Residential Energy Consumption Survey (RECS) conducted by the U.S. Department of Energy to measure energy consumption and equipment in housing units (HUs).[4] RECS is a multistage sample of counties, geographic segments within counties, and HUs within segments. In 2009, 430 counties, 3,000 segments, and about 12,000 HUs were sampled. RECS uses the Fay–BRR with 244 replicates and $\rho = 0.5$.

---

4. See http://www.eia.gov/consumption/residential/.

**Example 5.8: BRR example with grouped strata and PSUs.** The following code (in `ex.5.8_brr.grouped.do`) estimates the means per housing unit, their standard errors, and 95% confidence intervals for the means for total British thermal units (`TOTALBTU`) and total BTUs for: space heating (`TOTALBTUSPH`); air conditioning (`TOTALBTUCOL`); water heating (`TOTALBTUWTH`); refrigerators, appliances, electronics, and lighting (`TOTALBTURFG`); and miscellaneous uses (`TOTALBTUOTH`).

```
. use http://www.stata-press.com/data/svywt/recs2009, clear
. svyset [pw=NWEIGHT], vce(brr) brrw(brr_weight_1 - brr_weight_244) fay(0.5)
> dof(244) mse

. svy: mean TOTALBTU TOTALBTUSPH TOTALBTUCOL TOTALBTUWTH TOTALBTURFG TOTALBTUOTH
  (output omitted)
```

```
Survey: Mean estimation              Number of obs    =         12,083
                                     Population size = 113,616,229
                                     Replications     =            244
                                     Design df        =            244
```

|             | Mean      | BRR *<br>Std. Err. | [95% Conf. Interval] |          |
|------------:|----------:|-------------------:|---------------------:|---------:|
| TOTALBTU    | 89629.7   | 574.2534           | 88498.58             | 90760.83 |
| TOTALBTUSPH | 37192.28  | 344.1522           | 36514.39             | 37870.17 |
| TOTALBTUCOL | 5591.392  | 92.52901           | 5409.134             | 5773.649 |
| TOTALBTUWTH | 15872.68  | 147.471            | 15582.2              | 16163.15 |
| TOTALBTURFG | 4261.267  | 33.63474           | 4195.015             | 4327.518 |
| TOTALBTUOTH | 26712.09  | 224.9414           | 26269.02             | 27155.17 |

∎

As with the jackknife, any adjustments to the base weights in the full sample for coverage and nonresponse errors should be applied to each BRR replicate.

## 5.4.3   Bootstrap

The bootstrap is the most versatile of the replication variance estimators available in finite population sampling. Efron (1982) invented the bootstrap for nonsurvey applications. McCarthy and Snowden (1985) were the first to investigate its use for finite population estimation. Rao and Wu (1988) extended the method to multistage probability sampling for the case of with-replacement PSU sampling. PSUs can be selected with different probabilities, and sampling within PSUs can be done by any method that allows unbiased estimation of PSU totals as discussed for the ultimate cluster estimator in section 5.2. As in other cases, this bootstrap is applied when PSUs are actually selected without replacement. An advantage of the bootstrap over the jackknife and BRR is that it can be applied to almost any estimand (unlike the jackknife, which does not work for quantiles) and to any sample size greater than or equal to 2 (unlike BRR, which, strictly speaking, requires $m_h = 2$ with $h = 1, \ldots, H$ and $H \geq 2$). Also, neither design strata nor PSUs need to be collapsed as mentioned for the grouped jackknife and BRR procedures.

In the finite population bootstrap, $B$ stratified simple random samples of PSUs are selected with replacement. (Note that this is a true with-replacement subsample, unlike the variance estimator, where the WR sampling is assumed but not actually implemented.) Suppose that $\widetilde{m}_h$ PSUs are selected with-replacement from the $m_h$ sample PSUs in stratum $h$. The Rao–Wu weight adjustments for each sample unit $k$ within the initial sample PSUs ($k \in s_{hi}$) are

$$
\begin{aligned}
d_k^* &= d_k \left[ \left\{ 1 - \sqrt{\tfrac{\widetilde{m}_h}{(m_h - 1)}} \right\} + \sqrt{\tfrac{\widetilde{m}_h}{(m_h - 1)}} \tfrac{m_h}{\widetilde{m}_h} \widetilde{m}_{hi}^* \right] \\
&\equiv d_k B_{hi}
\end{aligned}
$$

where $B_{hi}$ is defined by the quantity in the brackets and $\widetilde{m}_{hi}^*$ is the number of times that PSU $hi$ was selected in a particular bootstrap sample. As in earlier sections, $d_k$ is the inverse of the selection probability for unit $k$, and $d_k^*$ is computed for units in all sample PSUs, not just those in the bootstrap sample. Provided that $\widetilde{m}_h \leq (m_h - 1)$, all such weights will be nonnegative, but not otherwise.

The simplest choice of $\widetilde{m}_h$ is $\widetilde{m}_h = m_h - 1$. In that case, the weight adjustments are always nonnegative and are

$$
d_k^* = d_k \frac{m_h}{m_h - 1} \widetilde{m}_{hi}^*
$$

If a PSU is not selected in the bootstrap sample, its bootstrap weight is 0; if it is selected once, the weight adjustment is $m_h / m_h - 1$; if twice, the adjustment is $2(m_h / m_h - 1)$; and so on. Adjustments for coverage and nonresponse are then applied to each bootstrap sample to form the final bootstrap analysis weights.

The bootstrap variance estimator has the same form as the standard BRR variance:

$$
v_{\text{boot}} \left( \widehat{\theta} \right) = B^{-1} \sum_{\alpha=1}^{B} \left( \widehat{\theta}_\alpha - \widehat{\theta}^* \right)^2
$$

With the `mse` option of `svyset`, the variance estimate is centered around the full-sample estimate. Without it, the average of the bootstrap estimates is used for $\widehat{\theta}^*$.

A decision in implementation is how many bootstrap samples to select. There is no clear theoretical guidance on choosing $B$. The theory for the bootstrap requires that $B \to \infty$, but, life being short, we make do with less than infinity. Gould and Pitblado (2017) recommend some experimentation with different values of $B$ to determine when bootstrap SEs become stable. In nonsurvey applications, analysts are usually limited only by how long they are willing to wait for their computer to do the bootstrapping. A limiting factor in survey samples with public-use files is that the number of weights per record is $B + 1$—one for the full-sample and $B$ bootstrap weights. In many applications, a few hundred replications are supplied. Statistics Canada (2005) lists the number of bootstrap replications used in several Statistics Canada surveys; in those surveys, $B$ was 100, 200, 500, 1,000, or 2,000.

**Example 5.9: Bootstrap example.** For the National Maternal and Infant Health Study (NMIHS), Gonzalez, Krauss, and Scott (1992) used bootstrap estimation for variances and is an example in StataCorp (2017).[5] The NMIHS was based on questionnaires administered to nationally representative samples of mothers with live births, stillbirths, and infant deaths during 1988 and to physicians, hospitals, and other medical care providers associated with those outcomes. The survey was based on a stratified, single-stage sample of live births, fetal deaths, and infant deaths.[6] The example below (see ex.5.9_bootstrap.nmihs.do) illustrates the svyset syntax for specifying the bootstrap weights using the bsrweight option. The mean birthweight of babies is estimated. Also, we save the 1,000 replicate estimates using the saving option in the svy: mean command. The default variable name for the replicate estimates will be the prefix _b_ plus the name of the variable for which the estimate is computed, but you are free to specify a name for the output variable if you prefer. All 1,000 replicate estimates can be inspected with list. The example then draws a histogram of the replicate estimates and saves the graph to an encapsulated postscript file (eps extension). As the histogram in figure 5.1 shows, the distribution of the estimated mean is symmetric, implying that a normal approximation confidence interval should work well in this case.

```
. use http://www.stata-press.com/data/svywt/nmihs_bs1000, clear
. svyset [pweight=finwgt], bsrweight(bsrw1 - bsrw1000) vce(bootstrap) mse

. svy, nodots saving("breps", replace): mean birthwgt
(note: file breps.dta not found)
Survey: Mean estimation            Number of obs     =       9,946
                                   Population size =   3,895,562
                                   Replications      =       1,000

                       Observed    Bstrap *
                          Mean     Std. Err.     [95% Conf. Interval]

          birthwgt     3355.452     6.53009       3342.654    3368.251


.          * load the replicate file of mean estimates
. use breps, clear
.          * list all 1000 reps
. list _b_birthwgt
. histogram _b_birthwgt
. graph export nmihs-bwt-histogram.eps, as(eps)
```

5. The data file with 1,000 bootstrap replicates is available at http://www.stata-press.com/data/r12/nmihs_bs.dta and is also one of the files that accompany this book.

6. See https://www.cdc.gov/nchs/nvss/nmihs.htm.

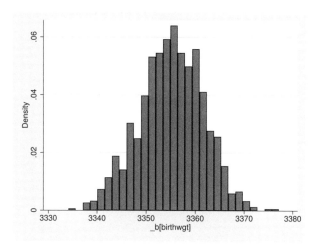

Figure 5.1. Histogram of 1,000 bootstrap estimates of birthweight from the NIMHS sample

■

Bootstrap weights can be generated using the `bsweights` command (Kolenikov 2010). The latest version of `bsweights` can be downloaded by typing `net from http://staskolenikov.net/stata` and then selecting `bsweights` from among several ado-files that Kolenikov has written. The example below uses the `smhosam.dta` sample that we used earlier in example 5.3 to illustrate the jackknife. Recall that this sample is an stsrswor of 30 hospitals selected from each of 4 strata.

**Example 5.10: Generating bootstrap replicate weights.** The `bsweights` command automatically reads the survey design characteristics that are currently `svyset`. `bsweights` requires a variable prefix and the options `n()`, and `reps()`. We use `bwt_` for the variable prefix, so the generated bootstrap replicate weight variables will be named `bwt_1`, `bwt_2`, `bwt_3`, .... The `n()` option specifies how to sample the PSUs within each stratum. We specify `n(raowu)` so that the PSUs are sampled according to Rao and Wu (1988). The `reps()` option specifies how many bootstrap replicate weight variables to generate. The `seed()` option of `bsweights` allows you to rerun this example and get the same results as presented here. The code for this example is in `ex.5.10_bootstrap.smho.do`.

```
. use dta/smhosam, clear
(Written by R.              )
. svyset ID [pweight=wt], strata(hosptype)

     pweight: wt
         VCE: linearized
 Single unit: missing
    Strata 1: hosptype
        SU 1: ID
       FPC 1: <zero>
. bsweights bwt_, n(raowu) reps(500) double seed(339487731)
. quietly svyset ID [pweight=wt], bsrweight(bwt_*) vce(bootstrap) mse
. svy, nodots: mean SEENCNT BEDS, cformat(%9.0gc)
Survey: Mean estimation              Number of obs   =        120
                                     Population size =        725
                                     Replications    =        500
```

|          | Observed<br>Mean | Bstrap *<br>Std. Err. | [95% Conf. Interval] | |
|----------|------------------|-----------------------|--------------|--------------|
| SEENCNT  | 1,669.17         | 193.8214              | 1,289.29     | 2,049.05     |
| BEDS     | 80.9263          | 8.15085               | 64.95093     | 96.90167     |

As in example 5.3, we compute the means of SEENCNT and BEDS, but this time, their SEs are estimated using the bootstrap replicates. The means are exactly the same as in the jackknife computation because the same full-sample weights are used. The bootstrap SEs are slightly larger than those of the jackknife.

■

## 5.4.4   Grouping PSUs to form replicates

To form replicates, strata or PSUs are sometimes grouped as mentioned for the jackknife in section 5.4.1. Two reasons that this is done are 1) to reduce the number of replicates for the jackknife or BRR methods (but not bootstrap) to a manageable size without limiting the analytic power (for example, degrees of freedom for variance estimation), and 2) to create pseudo- (or analytic) strata (VarStrat) when only one PSU has been selected per stratum or when only one PSU in a stratum participates. The methods for stratum-grouping are discussed in Valliant, Dever, and Kreuter (2013). Properly done, such grouping will still give legitimate variance estimators in that they are approximately unbiased and consistent.

In some sample designs, the number of PSUs may be extremely large so that the number of replicates for, say, the jackknife would be in the hundreds or thousands. This will often be the case in telephone surveys of persons or establishment surveys. Nonetheless, a database constructor may want to take advantage of replication to reflect the effects of multiple weighting steps. Grouping strata or PSUs is one solution to limiting the number of replicates that users must handle. Appendix D of the WesVar manual (Westat 2007) describes how the grouping can be done for several common sample designs. If you are using a database constructed by someone else (for example, a public-use file created by a government agency), there should be fields on the file for the strata and PSUs required for variance estimation. There is probably no choice but to assume that any grouping to create the strata and PSU variables was done in a statistically defensible way.

The grouped strata and PSU fields should be specified in `svyset` using the `strata()` option and the *psu* specifier. In example 5.2, the fields, `SDMVSTRA` and `SDMVPSU`, specified in `svyset` were grouped strata and PSUs.

**One PSU per stratum design**. Strata may be combined when one PSU is selected from each stratum. This is often done in area probability surveys in strata with PSUs that are not certainties. A 1-PSU-per-stratum design with strata formed based on geographic location forces the sample to be spread over the areas in the frame. If a 2-PSU-per-stratum design is used, the geographic spread could be less for a particular randomly selected sample with the number of PSUs fixed by the project budget. Strata in the 1-PSU design are usually paired for variance estimation with the field identifying the paired stratum being the `psu` value in `svyset`.

## 5.5 Effects of multiple weight adjustments

As shown in figure 1.2, surveys often have a series of weighting steps. Each step—applying base weights, adjusting for unknown eligibility, adjusting for nonresponse, and calibrating to population control totals—can affect variances. All the steps may help reduce bias. Depending on the application, variances may be increased or decreased by a particular step. As illustrated in Valliant (2004), accounting for each step when estimating variances can be important.

Reflecting the effects of these adjustments on variances is easier with some variance estimation methods than others. If the only steps are applying base weights and then using a simple calibration method, like poststratification, a linearization variance estimate that correctly accounts for these steps is available in Stata. If a nonresponse adjustment is added, then the linearization estimators in which the nonresponse-adjusted weights are treated as `pweights` do not fully account for the complexity that multiple steps introduce. This applies to variances computed by Stata and all other packages that handle survey data. On the other hand, replication variances can implicitly account for the effects of all steps. Any of the types of replication covered in section 5.4 are possibilities.

Consider a situation where a nonresponse adjustment is followed by calibration to control totals. The basic idea in using replication to reflect the effects on variances of both steps is to repeat the steps for every replicate. First, the full sample is divided into replicates based on which method will be used (jackknife, BRR, or bootstrap). By full sample, we mean both eligible respondents and nonrespondents because both are used in computing nonresponse adjustments. (If an unknown eligibility adjustment is made, respondents, nonrespondents, and units with unknown eligibility are all assigned to replicates.) The nonresponse adjustment is made for the full sample and separately for each replicate. Given the nonresponse-adjusted full-sample and replicate weights, the calibration is done for the full sample and for each replicate. This yields a full-sample weight and a set of replicate weights for each respondent that will implicitly reflect the combined effects of the steps. The procedure survwgt (Winter 2002) is one option for stacking the steps.

**Example 5.11: Accounting for nonresponse adjustment and calibration.** This example uses one of the nhis datasets, makes a nonresponse adjustment using cells, and rakes to margins on two variables, all using JKn replication. The full code is in ex.5.11_multsteps.survwgt.do. The input dataset was saved in example 3.2, where propensity classes were created for nonresponse adjustment. After some labeling of values of the two raking variables—age group and Hispanic—the survey weight, svywt, is divided by 1000 to reduce their sizes and create svywt1. The command

```
survwgt create jkn, psu(psu) strata(stratum) weight(svywt1) stem(JKn_)
```

treats svywt1 as a base weight and creates a set of JKn weights, JKn_1 – JKn_174. These are then used to create nonresponse-adjusted weights in

```
survwgt nonresponse [all], by(pclass) respvar(resp) stem(JKn_NR)
```

The variable specification [all] means that the nonresponse adjustment is made separately for the full-sample weights and each of the sets of JKn weights. The particular type of weight adjustment done in each pclass cell is to increase the weights of respondents so that they sum to the total of the weights for all sample units in the cell (respondents + nonrespondents).

In the final step, control totals are created for age_r1 and hisp. The population totals are constructed to reflect some undercoverage, that is, the sums of the nonresponse-adjusted weights in the categories of the raking variables are less than the control totals. The nonresponse-adjusted full-sample and replicate weights are then raked to the totals, age_tot and hisp_tot, in the command

```
survwgt rake [all], by(age_r1 hisp) totvars(age_tot hisp_tot) stem(JKNRrake)
```

This creates a series of JKn weights, JKNRrake0 – JKNRrake174. By default, survwgt labels the full-sample raked weight as JKNRrake0.

```
. use http://www.stata-press.com/data/svywt/nhis.logistnr.dta, clear
. recode age_r (3=3) (4=4) (5=5) (6=6) (7=6) (8=6), gen(age_r1)
. label define age_lab 3 "18-24" 4  "25-44" 5 "45-64" 6 "65+"
. label values age_r1 age_lab
. label define ed_lab 1 "High school or less" 2 "Some college"
> 3 "Bachelor or assoc" 4  "Master's & higher"
. label values educ_r ed_lab
. generate svywt1 = svywt/1000
```

```
.       * create JKn base weights using svywt1
. survwgt create jkn, psu(psu) strata(stratum) weight(svywt1) stem(JKn_)
.       * use pclass variable created in example~3.2 for NR adjustment cells
. survwgt nonresponse [all], by(pclass) respvar(resp) stem(JKn_NR)
.       * Rake to age_r1 and hisp margins
.       * Create age_r1 and hisp totals to illustrate undercoverage
. generate age_tot    = 1700
. replace age_tot = 5000 if age_r1 == 4
. replace age_tot = 4500 if age_r1 == 5
. replace age_tot = 2200 if age_r1 == 6
. generate hisp_tot    = 12060
. replace hisp_tot =  1340  if hisp == 2
. survwgt rake [all], by(age_r1 hisp) totvars(age_tot hisp_tot) stem(JKNRrake)

. svrtab educ_r hisp, count se format(%12.1f)
Cross-tabulation with replication-based (jkn) standard errors
Analysis weight:      JKNRrake0           Number of obs       =      2699
Replicate weights:    JKNRrake1...        Population size      =     13400
Number of replicates: 174                 Degrees of freedom  =        87
```

|                    |     | hisp    |         |         |
|-------------------:|:---:|--------:|--------:|--------:|
| educ_r             |  .  |       1 |       2 |   Total |
| High school or less |    |  8240.5 |   628.5 |  8869.0 |
|                    |     | (282.0) |  (23.8) | (287.4) |
| Some college       |     |  1669.9 |   239.0 |  1908.8 |
|                    |     | (177.1) |  (15.2) | (180.1) |
| Bachelor or assoc  |     |  1758.4 |   338.8 |  2097.2 |
|                    |     | (247.3) |  (16.9) | (250.6) |
| Master's & higher  |     |   391.2 |   133.7 |   524.9 |
|                    |     |  (98.3) |  (11.4) |  (99.9) |
| Total              |     | 12060.0 |  1340.0 | 13400.0 |
|                    |     |   (0.0) |   (0.0) |         |

```
  Key:  weighted counts
        (standard errors of weighted counts)
  Pearson:
    Uncorrected   chi2(3)          =    65.4948
    Design-based  F(2.70, 234.67) =    20.8219      P = 0.0000

. svyset psu [pw = JKNRrake0], strata(stratum)
```

```
. svy: tabulate educ_r hisp if resp==1, count se stubwidth(18)
(running tabulate on estimation sample)
Number of strata    =        87         Number of obs     =      2,699
Number of PSUs      =       174         Population size    =     13,400
                                        Design df         =         87
```

| educ_r | hisp 1 | hisp 2 | Total |
|---|---|---|---|
| High school or les | 8241 (706.9) | 628.5 (33.6) | 8869 (704.9) |
| Some college | 1670 (184.2) | 239 (14.75) | 1909 (186.7) |
| Bachelor or assoc | 1758 (292.5) | 338.8 (17.14) | 2097 (295) |
| Master's & higher | 391.2 (97.78) | 133.7 (11.73) | 524.9 (98.25) |
| Total | 1.2e+04 (865) | 1340 (43.53) | 1.3e+04 |

```
Key:  weighted count
      (linearized standard error of weighted count)

Pearson:
  Uncorrected   chi2(3)            =    65.4948
  Design-based  F(2.70, 234.48) =    20.6456       P = 0.0000
```

The first table above uses `svrtab`, which is part of the `svr` package, to estimate the counts of persons in an education × Hispanic table. SEs are shown in parentheses. For comparison, the second table above shows the same table with an `svyset` specification that does not account for the nonresponse adjustment or the raking. The point estimates of counts are the same because both tables use the same final full-sample weights. The SEs are similar except for "high school or less", where the SE for the marginal total is 287.4 when accounting for the NR adjustment and raking but 704.9 when ignoring them. The SEs for the Hispanic counts within "high school or less" are also much lower when accounting for both adjustments.

■

As noted above, we did account for variation in the nonresponse adjustment and the raking by redoing those steps separately for every replicate. However, these steps were conditional on the particular cell assignments (that is, `pclass` values) found in example 3.2. There is some randomness in those assignments because a logistic model was fit to estimate propensities of responding. In the example above, we could have stepped back and fit the propensity model separately in every replication. This would have created a set of `pclass` assignments that could, potentially, have been different from one replicate to another. This added source of variation might have increased SEs somewhat from those reported above.

The direction of the SE comparison in example 5.11 can go either way—accounting for nonresponse and calibration adjustments can either decrease or increase SEs. Effects of the adjustments also depend on the type of estimate and the analysis variable. Totals are likely to be most affected by the adjustments. The SEs of means and model parameter estimates, which are ratios, may be less affected. In any case, using a variance estimation method that properly accounts for the effects of the adjustments is important for obtaining more honest SEs.

One proviso in redoing adjustments for every replicate is that iterative procedures will not necessarily converge for every replicate even though they do for the full sample. If, for example, the number of units in a category of a raking variable is small in a replicate, the raking procedure may not be able to hit all control totals for that replicate. In that case, categories may have to be collapsed. If a dynamic procedure is used that allows the collapsing to potentially be different in the full sample and the replicates, then you will be reflecting that "procedural" variation. This is similar to reflecting uncertainty in model selection in other areas of statistics (for example, see Hoeting, Madigan, Raftery, and Volinsky [1999]). Checking the convergence of the iterative algorithms in every replicate is also an important quality control step (see chapter 8).

# 6   Nonprobability samples

Probability sampling became the touchstone for good survey practice decades ago after Neyman (1934) presented the theory for stratified and cluster sampling based on the randomization distribution. Neyman also showed that a type of nonrandom quota sample of Italian census records drawn by Gini and Galvani had failed to provide satisfactory estimates for many variables in the census. Quoting Smith (1976), "This combined attack was overwhelming and since that day random sampling has reigned supreme." Another early nail in the coffin of nonrandom sampling was the notable failure of one enormous, but nonprobability, sample to correctly forecast the 1936 U.S. presidential election result. In pre-election polls, the Literary Digest magazine collected 2.3 million mail surveys from mostly middle-to-upper income respondents. Although this sample size was huge, the poll incorrectly predicted that Alf Landon would win by a landslide over the incumbent, Franklin Roosevelt. In fact, Roosevelt won the election easily, carrying every state except for Maine and Vermont (Squire 1988). As Squire noted, the magazine's respondents consisted mostly of automobile and telephone owners plus the magazine's own subscribers. This pool underrepresented Roosevelt's core of lower-income supporters. In the same election, several pollsters (Gallup, Crossley, and Roper) using much smaller but more representative samples correctly predicted the outcome (Gosnell 1937).[1]

However, in many types of surveys using probability samples, response rates have declined dramatically, casting doubt on how well these samples represent the population. Pew Research reported that their response rates in typical telephone surveys dropped from 36% in 1997 to 9% in 2012 (Kohut et al. 2012). With such low response rates, a sample initially selected randomly can hardly be called a probability sample under a MAR assumption (see introduction to chapter 3). Low response rates raise the question of whether probability sampling is worthwhile, at least for some applications.

For some purposes, nonprobability samples have long been acceptable. For example, using convenience samples in experimental studies is standard practice, even when the conclusions are intended to apply to some larger population. The inferences are model-based and come from assumptions that the experimental effects are homogeneous among all units in the relevant population. Inferences from nonprobability survey samples must also rely on the adequacy of the models rather than the distribution generated by random sampling, to project a sample to a larger finite population.

---

1. However, it is worth noting that in the 1948 presidential elections, Gallup and Roper erroneously forecasted that Dewey would win using methods similar to those from 1936. The apparent reason was that they quit polling a week before the election and missed a late swing in voter sentiment (Lester 1998).

Obtaining data without exercising much control over the set of units for which it is collected is often cheaper and quicker than probability sampling, where efforts are made to use a frame that covers most of or all the population, and units are randomly selected from the frame. Because sampling frames are not always available, you may need to build the frame(s) prior to sample selection. Also, attempting to get nonrespondents to cooperate, which is standard procedure in probability samples, can be expensive and time-consuming. Capping the number of contacts is an expedient way of cutting costs. In telephone-only surveys, no amount of call attempts by interviewers is likely to boost response to the rates that were considered minimally acceptable 10 to 15 years ago. For these reasons, nonprobability sampling is currently staging a kind of renascence (for example, see Dever and Valliant [2014]).

There is much controversy about using nonprobability surveys for making inferences to a population. The American Association for Public Opinion Research (AAPOR) has issued two task force reports on using these samples—neither of which favored their use. Baker et al. (2010) studied using online Internet panels; Baker et al. (2013a,b) cover nonprobability sampling generally. Nonprobability surveys capture participants through various methods. The AAPOR task force on nonprobability sampling (Baker et al. 2013a) characterized these samples into three broad types:

1. Convenience sampling
2. Sample matching
3. Network sampling

Baker et al. (2013a) describe these in some detail; we briefly summarize them here.

## Convenience sampling

Convenience sampling is a form of nonprobability sampling in which researchers primarily consider easily locating and recruiting participants. The convenience is for the researchers; no formal sample design is used. Some types of convenience samples are mall intercepts, volunteer samples, river samples, and observational studies. In a small intercept sample, interviewers try to recruit shoppers to take part in some study. Usually, neither the malls nor the people are probability samples.

Volunteer samples are common in social science, medicine, and market research. Volunteers may participate in a single study or become part of a panel in which members may be recruited for different studies over the course of time. A more recent development is the opt-in web panel, in which volunteers are recruited when they visit particular web sites. We address methods for weighting these panels below. After becoming part of a panel, the members may participate in many different surveys, often for some type of incentive. River samples are a version of opt-in web sampling in which volunteers are recruited at a number of websites for one or more (panel) surveys. Some thought may be given to the set of websites used for recruitment with an eye toward obtaining a cross-section of demographic groups or a special group of interest (for example, adults aged 18–29 who smoke).

One extreme case of a convenience sample used to accurately predict the outcome of the 2012 U.S. presidential election was of Xbox gaming console users (Wang et al. 2015). The key in that study was using an extremely elaborate estimator, which we describe in more detail below. Favorable results such as this encourage belief that probability sampling could be jettisoned entirely. However, as noted earlier, there are famous examples of extremely large nonprobability samples with results that were proved to be completely wrong. More recent examples of polls that failed to correctly predict election outcomes are the 2015 British parliamentary election[2], the 2015 Israeli Knesset election[3], and the 2014 governor's race in the United States state of Maryland[4]. Many United States polls incorrectly projected the winner of the 2016 presidential race, especially at the state level. AAPOR convened a panel of survey research and election polling experts to conduct a post-hoc analysis of the 2016 polls.[5]

There were various reasons for the misfires, including samples with low contact and participation rates, samples based on unrepresentative volunteer panels, and volatility in voters' opinions about candidates. The Market Research Society and British Polling Council of Britain (Sturgis et al. 2016) concluded that unrepresentative samples with defects that could not be corrected through weighting were the main culprits in the failure of the 2015 British parliamentary polls. In any case, the methods of obtaining public opinion and of estimation used by most polling firms failed to correctly predict the outcomes in all the examples above.

Observational studies are often used as a substitute for randomized experiments when investigating the effects of medical or social phenomena. These studies are especially convenient when trying to determine whether some factor causes, or at least is related to, a particular health condition. In some situations, randomized experiments are impossible, while an observational study is not. For example, randomizing persons into smoking and nonsmoking groups to decide whether smoking cigarettes is related to developing lung cancer would be unethical, but finding a collection of people who have smoked and comparing them with nonsmokers is feasible. Techniques developed by Rosenbaum and Rubin (1983) and others for analyzing such observational data have been applied when attempting to develop weights for some volunteer samples. This topic is discussed in detail below.

## Sample matching

In sample matching, the members of a nonprobability sample are selected to match a set of important population characteristics. For example, a sample of persons may be constructed so that its distribution by age, other race, and sex closely matches the distribution of the inference population. Quota sampling is an example of sample

---

2. See BBC News, http://www.bbc.com/news/uk-politics-32751993.
3. See J Street, http://www.israelelection2015.org/polls/.
4. See FiveThirtyEight DataLab
   http://fivethirtyeight.com/datalab/governor-maryland-surprise-brown-hogan/.
5. See https://www.aapor.org/Publications-Media/Press-Releases/AAPOR-to-Examine-2016-Presidential-Election-Pollin.aspx.

matching. The matching is intended to reduce selection biases if the covariates that predict survey responses can be used in matching. Rubin (1979) presents the theory for matching in observational studies.

A variation of matching in survey sampling is to match the units in a nonprobability sample with those in a probability sample. Each unit in the nonprobability sample is then assigned the weight of its match in the probability sample. Rivers (2007) describes this type of sample matching in the context of web survey panels. Propensity scores can also be used for matching by using methods similar to those for nonresponse adjustment (see section 3.2). The nonprobability and probability samples can be combined, and the propensity of being in the nonprobability sample can be estimated with logistic regression. All units (nonprobability and probability) are sorted by their propensities. The nearest probability sample neighbor to each nonprobability unit is found, and its weight is assigned to the nonprobability unit. Stuart et al. (2011) similarly use propensity cores to generalize results from randomized trials to populations. Note that calibration may still be necessary to scale the weights for population estimation after doing propensity matching.

## Network sampling

In network sampling, members of some target population (usually a rare one like intravenous drug users or men who have sex with men for money) are asked to identify other members of the population with whom they are somehow connected, for example, via a social network. Members of the population that are identified in this way are then asked to join the sample. This method of recruitment may proceed for several rounds. Snowball sampling (also called chain sampling, chain-referral sampling, referral sampling, or link tracing) is an example of network sampling in which existing study subjects name or even recruit additional subjects from among their acquaintances. These samples typically do not represent any well-defined target population, although they are a way to accumulate a sizable collection of units from a rare population.

Sirken (1970) is one of the earliest examples of network or multiplicity sampling in which the network that respondents report about is clearly defined (for example, members of a person's extended family). Properly done, a multiplicity sample is a probability sample, because a person's network of recruits is well-defined. Heckathorn (1997) proposed an extension to this called "respondent driven sampling", in which an initial set of persons identified for the study (seeds) would report how many people they knew in a rare population (like intravenous drug users) and recruit other members of the rare population (see also Gile and Handcock [2015]). Note that the initial set of study participants (or nodes) can be a probability or a nonprobability sample (Thompson and Seber 1996; Thompson 2012). If some restrictive assumptions on how the recruiting is done are satisfied, inferences to the full rare population are possible, but these assumptions can easily be violated (for example, see Gile and Handcock [2010] and Gile, Johnston, and Salganik [2015]). Because the network applications are extremely specialized, we will not address them further.

In the rest of this chapter, we briefly review the weighting methods that can be used in volunteer sampling and statistical matching.

## 6.1   Volunteer web surveys

Because volunteer web surveys are one of the most common kinds of nonprobability samples, we emphasize them in this discussion, although the statistical estimation methods below can be applied in other situations also. Couper (2000) lists three kinds of nonprobability web surveys: 1) polls as entertainment, 2) unrestricted self-selected surveys, and 3) volunteer opt-in panels. Sites that are dedicated to polling for entertainment, such as www.misterpoll.com or survey.net, usually make no pretense of being scientific or of representing the opinions of anyone other than the people who participated in one of the polls.

Unrestricted self-selected surveys use open invitations on portals, frequently visited websites, or dedicated "survey" sites to do one-time surveys. Typically, these surveys have no access restrictions and sometimes do not control how many times a person completes a survey. Some sites even encourage it—the official site of U.S. major league baseball lets fans vote up to 35 times to pick players for the annual all-star game.[6] Other opt-in web surveys employ a series of checks to limit enrollment per person to a single occurrence (for example, see Brown, Dever, Squiers, and Augustson [2016]).

The third type of nonprobability web survey creates a volunteer panel by wide appeals on popular websites and portals. Demographic and other baseline information is collected from the volunteers at the time they register. Subsamples are then selected from the large panel and asked to participate in particular surveys. An example is a panel of millions of persons maintained by Harris Interactive.[7] Harris can mount online surveys in over 200 countries, including the United States, most countries in the European Union, Canada, China, India, and countries in South America. As claimed in their advertising material, "Although all respondents are Internet users, our ongoing parallel studies and extensive experience enable us to weight the sample to represent the adult population in most markets" (Harris Interactive 2014).[8]

Though some are general purpose, other nonprobability panels are tailored toward specific populations. For example, the U.S. Centers for Disease  Control (CDC) surveys a subsample from an opt-in panel maintained by Survey Sampling International to assess early-season influenza vaccination rates among pregnant women to enhance educational materials and public outreach (Ding et al. 2015).[9]  The CDC also assesses

---

6. See https://www.mlb.com.

7. See http://www.theharrispoll.com.

8. In contrast, some panels are formed by selecting an initial probability sample and recruiting persons to participate. For example, in the United States, GfK (2013) selects an initial address-based sample (Iannacchione 2011), which, in principle, covers almost the entire population, not just persons that use the Internet.

9. See https://www.surveysampling.com/contact/join-a-panel/.

vaccination rates for healthcare providers of certain disciplines by way of a sample of Medscape members, a professional health website maintained by WebMD Professional Services.[10, 11]

## Problems with opt-in panels

Nonprobability sampling is considered by some as a cost-effective alternative to more expensive probability-based methods (Couper 2013). Aside from some cost for maintenance, survey panels enable quick-targeted sampling using information obtained (and updated periodically) during recruitment, in contrast to sampling for a single survey. At first glance, the marriage of lower cost and quick turnaround seems the best way to go. Yet, there are many potential problems with opt-in web panels that are discussed by Baker et al. (2010). We sketch some of them here.

**Coverage and selection bias**. Coverage is a major problem—only persons with access to the Internet can complete a survey or join a panel. To describe three components of coverage survey bias, Valliant and Dever (2011) defined three populations, reproduced in figure 6.1: 1) the target population of interest for the study $U$ (universe); 2) the potentially covered population available given the way that the survey is done, $F_{pc}$; and 3) the actual covered population, $F_c$, and the portion of $F_{pc}$ that is recruited for the study through the essential survey conditions. For example, consider an opt-in web survey on smoking cessation treatments. The target population $U$ may be defined as adults aged 18–29 who currently use regular or electronic cigarettes. The potentially covered population $F_{pc}$ would be those study-eligible individuals with Internet access; those actually covered, $F_c$, would be the subset of the potential covered population who visit the sites where study recruitment occurs. This last problem is selection bias. The sample $s$ are those persons who are invited to participate in the survey and who actually do. The $U - F_{pc}$ area in the figure are the many persons who have Internet access but never visit the recruiting websites or who do not have Internet access at all. In most situations, $U - F_{pc}$ is vastly larger than $F_c$ or $F_{pc}$.

---

10. See http://www.medscape.com/public/about.
11. See https://www.cdc.gov/flu/fluvaxview/hcp-ips-nov2016.htm.

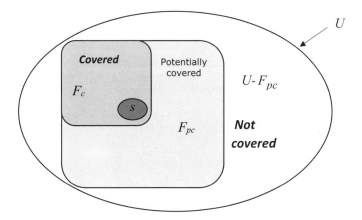

Figure 6.1. Illustration of potential and actual coverage of a target population

Table 6.1 shows the percentages of persons in European Union countries that have the "capacity" to use the Internet, meaning that a person must have available access to an Internet connection point and must have the basic knowledge required to use web technology. The percentages in the table correspond to the proportions that $F_{pc}$ is of the universe in a general population survey. The potentially covered population ranges from 96% in Denmark and Netherlands to 56% in Romania. Thus, coverage problems would range from minimal to substantial depending on the country.

Table 6.1. Percentages of persons in European Union countries with "capacity" to use the Internet in 2016.

| Country | Percent | Country | Percent |
|---|---|---|---|
| Denmark | 96 | Latvia | 82 |
| Netherlands | 96 | Czech Republic | 80 |
| Cyprus | 95 | Spain | 77 |
| Luxembourg | 95 | Hungary | 76 |
| Sweden | 95 | Croatia | 75 |
| Finland | 94 | Malta | 73 |
| United Kingdom | 92 | Slovenia | 73 |
| Germany | 88 | Portugal | 68 |
| Belgium | 85 | Poland | 68 |
| Estonia | 84 | Greece | 63 |
| France | 84 | Italy | 62 |
| Austria | 83 | Bulgaria | 57 |
| Slovakia | 83 | Romania | 56 |
| Ireland | 83 | | |
| Lithuania | 82 | Total European Union | 79 |

Source: http://www.internetworldstats.com/stats9.htm

Coverage problems also exist in the United States. Table 6.2 lists percentages of households in the United States in 2013 estimated from the American Community Survey (ACS) that have some type of Internet subscription (File and Ryan 2014).[12] The ACS estimates are based on a sample of about 3.5 million households. About 25% of households had no Internet access, which in itself is a substantial amount of undercoverage of the full population. The coverage varies considerably by demographic group. Only 58.3% of households where the head is 65 or older have the Internet. Black non-Hispanic and Hispanic households are less likely to have access than other races and ethnicities. Households in metropolitan areas are more likely to have access. There is also a clear dependence on income and education; as income and education increase, so does the percentage of households with access. As illustrated in Dever, Rafferty, and Valliant (2008), these coverage errors can lead to biased estimates for many items.

---

12. See www.census.gov/programs-surveys/acs/.

Table 6.2. Percentages of U.S. households with Internet subscriptions; 2013 American Community Survey

| Characteristics | Percent of households with some Internet subscription |
|---|---|
| Total households | 74.4 |
| Age of householder | |
| 15–34 years | 77.7 |
| 35–44 years | 82.5 |
| 45–64 years | 78.7 |
| 65 years and older | 58.3 |
| | |
| Race and Hispanic origin of householder | |
| White alone, non-Hispanic | 77.4 |
| Black alone, non-Hispanic | 61.3 |
| Asian alone, non-Hispanic | 86.6 |
| Hispanic (of any race) | 66.7 |
| | |
| Limited English-speaking household | |
| No | 75.5 |
| Yes | 51.4 |
| | |
| Metropolitan status | |
| Metropolitan area | 76.1 |
| Nonmetropolitan area | 64.8 |
| | |
| Household income | |
| Less than $25,000 | 48.4 |
| $25,000–$49,999 | 69.0 |
| $50,000–$99,999 | 84.9 |
| $100,000–$149,999 | 92.7 |
| $150,000 and more | 94.9 |
| | |
| Educational attainment of householder | |
| Less than high school graduate | 43.8 |
| High school graduate | 62.9 |
| Some college or associate's degree | 79.2 |
| Bachelor's degree or higher | 90.1 |

Selection bias occurs when some groups in the potentially covered population $F_{pc}$ are more likely to volunteer for an opt-in web survey. Bethlehem (2010) reviews this issue for web surveys. Willems, van Ossenbruggen, and Vonk (2006) report that ethnic minorities and immigrant groups were systematically underrepresented in Dutch panels. They also found that, relative to the general population, the Dutch online panels contained disproportionately more voters, more Socialist Party supporters, more heavy Internet users, and fewer churchgoers. As noted by Kennedy et al. (2016), estimates from panels can vary widely so understanding the coverage properties is important—this information is hard to discover.

**Nonresponse** of several kinds affects opt-in web surveys, both one-time studies and web panels. People may click on a banner ad advertising the survey or panel but never complete all registration steps. Alvarez, Sherman, and Van Beselaere (2003) report that during the recruitment of one panel, just over 6% of those who clicked through a banner ad to the panel registration page eventually completed all the steps required to become a panel member. Squiers et al. (2016) report that among the almost 154,000 visitors to multiple survey recruitment pages for a smoking cessation study, approximately 17.8% began the screener, owing to a likely combination of nonresponse and nonsmoking status. Many studies, however, do not collect such information, so the recruitment rates are not reported.

Many panel vendors and some researchers using online surveys have a "double opt-in" procedure for joining a panel. First, a person registers his or her name, email, and some demographics. Then, the vendor sends the person an email that must be responded to in order to officially join the panel. This eliminates people who give bogus emails and those enrolling multiple times with the same email (presumably to increase their take of the incentive pool) but also introduces the possibility of "registration nonresponse" because some people do not respond to the vendor's email.

Finally, a panel member asked to participate in a survey may not respond. Baseline information (much like with frame data) may be useful in understanding and correcting for nonresponse patterns in the survey.

**Attrition** is another problem—persons may lose interest and drop out of a panel. Many surveys are targeted at specific groups, for example, young black females. A panelist that is in one of these "interesting" groups may be peppered with survey requests and drop out for that reason. Another reason that some groups, like the elderly, are overburdened is that they may be oversampled to make up for anticipated nonresponse. Vendors of opt-in panels, like probability-based panels, employ a series of procedures to maintain and sometimes refresh the panel with new members.

**Measurement error** is also a worry in these surveys. The types of error that have been demonstrated in some studies are effects that are largely due to questionnaire design, mode, and peculiarities of respondents. For example, the persons who participate in panels tend to have higher education levels. The motivation for participating may be a sense of altruism for some but may be just to collect an incentive for others. Participants are often paid per survey completed. Some respondents speed through surveys, answering as quickly as possible to collect the incentive. This is a form of "satisficing" where respondents do just enough to get the job done (Simon 1956). On the other hand, self-administered online surveys do tend to elicit more reports of socially undesirable behaviors, like drug use, than do face-to-face surveys (Kreuter, Presser, and Tourangeau 2008). Higher reports are usually taken to be more nearly correct. But, it may be that the people taking those surveys just behave undesirably more often than the general population.

Baker et al. (2010, 739) list 19 studies where the same questionnaire was administered by interviewers to probability samples and online to nonprobability samples. As they noted, "Only one of these studies yielded consistently equivalent findings across methods, and many found differences in the distributions of answers to both demographic and substantive questions. Further, these differences generally were not substantially reduced by weighting." The same can be said for the 8 studies evaluated by Tourangeau, Conrad, and Couper (2013) and other studies too numerous to name here.

Despite these actual and potential problems, opt-in web surveys and online panels are now widely used. Baker et al. (2010) quote the market research newsletter *Inside Research* as estimating the total spent on online research in 2009 at about $2 billion USD, the vast majority of which is supported by online panels. There are a few approaches that have been developed for weighting these nonprobability samples, which we cover below.

## 6.2 Weighting nonprobability samples

Baker et al. (2013a) discuss the proposed methods for weighting nonprobability samples. Such samples lack many of the features that guide weighting in probability samples. A nonprobability sample is not selected randomly from an explicit sampling frame. Consequently, selection probabilities cannot be computed, and the usual method of computing base weights (inverse of the selection probabilities) covered in chapter 2 does not apply. In fact, the repeated sampling analysis that underlies design-based inference does not apply. As a result, another mathematical approach must be used to derive statistical properties.

In any inference problem, the standard approaches involve averaging over some random process to calculate biases, variances, and other statistical properties. In design-based inference, the random process is the drawing of the sample itself, which is controlled by the sampler. Because this method is inapplicable for nonprobability samples, models are used. One option is to model the mechanism by which units appear in the sample as being random (even though the mechanism is not controlled by the sampler). Pseudo-inclusion probabilities are estimated and used to construct weights. This might be called "quasirandomization-based inference". Another option is to model the variables collected in the survey as being functions of covariates. A statistical model is then used to project the sample to a target population. This is an example of model-based or prediction inference discussed in chapter 1 (see also Valliant, Dorfman, and Royall [2000]).

In the remainder of this section, we assume that weights are needed to allow estimation of population totals. These weights can also be used for estimating means, proportions, and model parameters but requiring that the weights be scaled for totals puts them on the same footing as the weights discussed previously for probability samples. If only means, proportions, and other quantities that are essentially ratios were estimated, the scaling would not be important. But, as noted in section 7.1, we recommend creating weights that are appropriate for estimating population totals, because these weights are generally more useful.

## Quasirandomization weighting

In the quasirandomization approach, pseudo-inclusion probabilities are estimated and used to inflate the sample to a full population. Exactly what the "full population" is will be determined by the way in which the pseudo-inclusion probabilities are estimated, as described below. In any case, the resulting weights are intended to correct for any coverage, selection, and nonresponse errors in the sample. Given estimates of the pseudoprobabilities, design-based formulas can be used for point estimates and variances. To illustrate how involved estimating these probabilities may be, consider a case in which a volunteer panel of persons is recruited to provide a pool from which a sample of persons is selected. To respond to a survey, a person must have Internet access, volunteer for the panel, be selected for the particular survey, and then respond. We denote the set of persons with access to the Internet by $W$ and the set of volunteers by $V$. Considering all of these, the probability of person $i$ participating in that web survey can be decomposed as

$$\pi\left(\mathbf{x}_i\right) = \pi\left(i \in W \,|\mathbf{x}_i\right) \pi\left(i \in V \,|W, \mathbf{x}_i\right) \pi\left(i \in s_V \,|V, W, \mathbf{x}_i\right) \pi\left(i \in s_{VR} \,|s_V, V, W, \mathbf{x}_i\right)$$
$$(6.1)$$

where

$\mathbf{x}_i$ = a vector of covariates for person $i$ that are predictive of participation;

$\pi\left(i \in W \,|\mathbf{x}_i\right)$ = probability of having access to the Internet;

$\pi\left(i \in V \,|W, \mathbf{x}_i\right)$ = probability of volunteering for an opt-in panel given that person $i$ has access to the Internet;

$\pi\left(i \in s_V \,|V, W, \mathbf{x}_i\right)$ = probability that person $i$ was subsampled from the panel and asked to participate with $s_V$ denoting the subsample from the panel;

$\pi\left(i \in s_{VR} \,|s_V, V, W, \mathbf{x}_i\right)$ = probability that person $i$ responds given selection for the subsample with $s_{VR}$ denoting the set of survey respondents.

The standard methods from earlier chapters can be used to compute the last two terms in (6.1). The first two probabilities—having Internet access and volunteering for the panel—are more difficult. Both are likely to depend on the $\mathbf{x}_i$ covariates. For example, persons with higher socioeconomic status are more likely to have Internet access; younger people are more likely to join a panel than older ones. In some countries, probability samples that represent the full population may include questions on Internet access. For example, the U.S. National Health Interview Survey routinely includes such questions. The probability of volunteering (given Internet access) is harder to estimate.

**Reference survey**. One approach to estimating

$$\pi\left(i \in W \mid \mathbf{x}_i\right) \times \pi \ \left(i \in V \mid W, \mathbf{x}_i\right)$$

for either an opt-in, one-time web survey or an online panel is to use a reference survey in parallel to the nonprobability survey. The reference survey is a probability survey that can be selected from one of the following:

1. the population of persons who have Internet access or

2. the full population including persons that do not have the Internet.

A key consideration is the following:

What population does the reference sample represent (or attempt to represent)?

The answer may depend on how survey weights were computed for the reference sample. For example, if the reference sample is 1) selected only from persons who have Internet access and the weights inflate that sample only to the Internet population, then the method described below will yield pseudo-inclusion probabilities of being in the non-probability sample within the Internet population. On the other hand, if weights for the reference sample 2) inflate it to the full population, including persons without the Internet, then we would have pseudo-inclusion probabilities of being in the nonprobability sample within the full population. (In either case, the reference sample weights should correct any undercoverage error or nonresponse bias within whatever population the reference survey purports to represent.)

The statistical approach is to combine the reference sample respondents and the sample of volunteers, code the units in the reference sample as 0 and the units in the nonprobability sample as 1, and fit a binary regression model to predict the probability of being a volunteer.[13] If the survey weights for the reference survey are used and the units in the nonprobability sample are all assigned a weight of 1, the estimated propensities will be of the probability of being in the nonprobability sample within whatever population the reference sample represents. If choice 1 above is used, then the estimated probability is that of volunteering given that a person has Internet access, that is, $\pi\left(i \in V \mid W, \mathbf{x}_i\right)$ in expression 6.1. If choice 2 is used, then an estimated propensity is that of the probability of having Internet access times the probability of volunteering given access, that is, $\pi\left(i \in W \mid \mathbf{x}_i\right)\pi\left(i \in V \mid W, \mathbf{x}_i\right)$ in (6.1), which is the probability within the full population regardless of Internet access.

---

13. Strictly speaking, the reference survey should not overlap with the volunteer sample at all, because persons in the reference are treated as nonvolunteers when fitting a binary regression (Valliant and Dever 2011). However, the proportion of the population that volunteers for a panel is so low that overlap is usually not a concern.

Using a reference survey to estimate volunteering probabilities is often followed by calibration to population counts of some kind (see chapter 4). If the reference survey is selected from persons who have Internet access as in item 1 above, there might be some ambiguity and confusion about what population the sample is supposed to represent. When the probability of volunteering given Internet access, $\pi (i \in V | W, \mathbf{x}_i)$, is estimated and its inverse is used as a weight, the represented population consists of those who have Internet access. But, if the calibration controls are for the full population regardless of Internet access, then the calibration serves as a correction for the lack of coverage of non-Internet persons. As described in previous chapters, the calibration can also function as a variance reducer.

The option used by some commercial marketing companies is to recruit a large panel, select subsamples from it, and have a parallel reference survey (for example, see Schonlau et al. [2004]). The recruitment may be done by telephone. Thus, the frame coverage can differ depending on the country. In countries where telephone coverage is almost complete, a telephone frame may correspond closely to option 2 above. However, telephone surveys suffer from high nonresponse, which can introduce its own coverage problems. Telephone surveys normally require calibration to external controls to correct both coverage and nonresponse problems.

A key requirement of the reference survey is that it includes the same covariates as the volunteer survey, so a binary regression can be fit to permit estimation of inclusion probabilities for the volunteers. One possibility for a reference survey is to use a publicly available dataset collected in a well-designed and executed probability survey (for example, by a central government agency). Another possibility is for the survey organization to conduct its own reference survey. In the latter case, some specialized questions, beyond the usual age, race, sex, or education demographics, can be added that are felt to be predictive of volunteering and of the analysis variables for the volunteer survey. Schonlau, van Soest, and Kapteyn (2007) refer to these extra covariates as "webographics". However, identifying webographics that are useful beyond the standard demographics (age, other race, sex, income, and education) is difficult (Lee and Valliant 2009). Of course, a problem with conducting your own reference survey is that doing a high-quality survey with good coverage of the target population is expensive and time consuming and may be beyond the means of many organizations.

Another important assumption for this methodology is known as "common support". For each value of the covariates $\mathbf{x}$ that is present in the population, the probability of being in the nonprobability sample or in the reference sample are both positive. That is, every unit in the population has some chance of showing up in either type of sample. This is analogous to the requirement in the probability sample that every unit have a nonzero selection probability. The common support requirement is also used in observational studies where every unit is assumed to have a positive probability of being "treated" or "not treated" (Rosenbaum and Rubin 1983). If, for example, some units have no chance of being observed in the nonprobability sample, then we cannot claim that the quasirandomization process represents the full population.

**Example 6.1: Pseudo-inclusion probabilities with an opt-in web survey.** To illustrate the estimation of pseudo-inclusion probabilities, we use a dataset derived from the 2003 Behavioral Risk Factor Surveillance Survey (BRFSS) in the state of Michigan (United States) originally used in Valliant and Dever (2011). The full set of code is in `ex.6.1_pseudoprobs.do`. The `internet.dta` file is a sample of 200 persons who have access to the Internet at home; this is the file that requires pseudoweights. `refpop.dta` is a reference population of 20,000 that covers both persons that have the Internet and who do not. Each file is loaded by the Stata program and an indicator variable, `in_internet`, is created for whether a person is in the opt-in Internet sample. The Internet file with the indicator is saved as `internet1.dta`; the reference file with its indicator is saved as `internet0.dta`. The two files are then combined with the `append` command.

We fit a logistic regression predicting whether a person is in the volunteer (nonprobability) Internet sample using age ($< 55$ or $\geq 55$), education level, and income. The inverses of the predicted probabilities for the Internet sample are saved as weights in `pseudowt`. These range from 49.5 to 658.8 and sum to 20,013.3, that is, about the size of the reference population. Note that if the reference dataset had been a probability sample with weights, then a weighted logistic regression would be run. The weights for the reference sample cases would be their survey weights, while the Internet sample cases would be set to 1. (See Valliant and Dever [2011] for a discussion of the reference survey weights.)

As a last step, the proportions of persons reporting different levels of general health status are estimated using the volunteer sample only. The `svyset [pweight=pseudowt]` command defines the weights for the Internet sample to be the inverse of the predicted logistic probabilities. Standard errors are estimated using the formula for with-replacement, varying probability sampling given in the special case of (5.3) with no strata and a sampling fraction of 0. We discuss the thinking behind this in section 6.3.

```
        . * create 0-1 indicators for being in Internet sample
. use http://www.stata-press.com/data/svywt/internet

. generate in_internet = 1

. save "internet1.dta"

. use http://www.stata-press.com/data/svywt/refpop, clear

. generate in_internet = 0

. save "internet0.dta"

. append using "internet1.dta"

        . * create indicator for age < 55
. generate newage = (AGECAT < 5)

. label define intlabel 0 "in reference sample" 1 "in Internet sample", replace

. label define edlabel 1 "1: Did not grad high school" 2 "High sch grad"
> 3 "Attended coll or tech sch" 4 "College or tech sch grad", replace

. label define agelabel 1 "18-24" 2 "25-34" 3 "35-44" 4 "45-54" 5 "55-64"
> 6 "65+", replace

. label define newagelabel 0 "< 55" 1 ">= 55", replace
```

```
. label define inclabel 1 "Less than $15K" 2 "$15K to less than $25K"
> 3 "$25K to less than $35K" 4 "$35K to less than $50K" 5 "$50K or more", replace
. label define hlthlabel 1 "Excellent" 2  "Very good" 3  "Good" 4 "Fair" 5 "Poor"
. label values in_internet intlabel
. label values AGECAT agelabel
. label values newage newagelabel
. label values EDCAT edlabel
. label values INCOMC3 inclabel
. label values GENHLTH hlthlabel
. logit in_internet i.newage i.EDCAT i.INCOMC3
```
  (*output omitted*)

| in_internet | Coef. | Std. Err. | z | P>|z| | [95% Conf. Interval] | |
|---|---|---|---|---|---|---|
| newage | | | | | | |
| >= 55 | .358074 | .1640902 | 2.18 | 0.029 | .0364631 | .6796848 |
| | | | | | | |
| EDCAT | | | | | | |
| High sch g.. | .5033889 | .4082349 | 1.23 | 0.218 | -.2967368 | 1.303515 |
| Attended c.. | .9144392 | .4009749 | 2.28 | 0.023 | .1285428 | 1.700336 |
| College or.. | .6283427 | .4084794 | 1.54 | 0.124 | -.1722622 | 1.428948 |
| | | | | | | |
| INCOMC3 | | | | | | |
| $15K to le.. | .7581934 | .3464601 | 2.19 | 0.029 | .079144 | 1.437243 |
| $25K to le.. | 1.079613 | .34522 | 3.13 | 0.002 | .4029943 | 1.756232 |
| $35K to le.. | 1.334336 | .3409602 | 3.91 | 0.000 | .6660667 | 2.002606 |
| $50K or more | 1.21109 | .3447048 | 3.51 | 0.000 | .5354813 | 1.886699 |
| | | | | | | |
| _cons | -6.488865 | .4667437 | -13.90 | 0.000 | -7.403666 | -5.574064 |

```
.     * create logistic predicted probs and pseudoweights
. predict pseudoprob, pr
. drop if in_internet==0
. generate pseudowt = 1/pseudoprob

. summarize pseudowt
```

| Variable | Obs | Mean | Std. Dev. | Min | Max |
|---|---|---|---|---|---|
| pseudowt | 200 | 100.0667 | 66.18541 | 49.5219 | 658.7762 |

```
. svyset [pweight=pseudowt]
```

```
. svy: tabulate GENHLTH, se stubwidth(10)
(running tabulate on estimation sample)
Number of strata    =         1          Number of obs     =        200
Number of PSUs      =       200          Population size   = 20,013.344
                                         Design df         =        199
```

| GENHLTH | proportion | se |
|---|---|---|
| Excellent | .1577 | .0262 |
| Very good | .3811 | .0401 |
| Good | .3271 | .0391 |
| Fair | .0814 | .0231 |
| Poor | .0527 | .035 |
| Total | 1 | |

```
Key:  proportion  =  cell proportion
        se        =  linearized standard error of cell proportion

. save "brf.internet1.wtd.dta", replace
```

In the final command, we save the file of nonprobability cases with their weights as brf.internet1.wtd.dta for later use in example 6.4.

∎

**Warning**. The nonprobability sample to be weighted must contain cases that cover all values of the required covariates to fit the propensity model. If, for example, African–American women 70 years and older have a much different propensity of participating than other demographic groups, and that group is not represented in the sample, the propensity will have to be estimated from some larger group, like all women over 70. Omission of a subgroup like this from the nonprobability sample naturally means that estimates cannot be made for the subgroup and also will typically lead to biased full-population estimates. This is an example where the common support assumption was violated.

Elliott (2009) presented an alternative to the above procedure. His method does require separately modeling the probability distribution of the covariates within the probability sample and the volunteer sample. As a result, it seems harder to implement in practice.

**Sample matching** is another attempt to reduce selection biases in a nonprobability sample. As noted in Baker et al. (2013a), the matching can be done on an individual or aggregate level. An example of individual-level matching is if, for each case in a volunteer sample, a matched case is found in a probability reference sample. The matches would

be found based on covariates available in each dataset. This may be done based on individual covariate values or on propensity scores as described in Rosenbaum and Rubin
(1983). Measures of closeness to signify a match are discussed in the next example.

Matching at the aggregate level is making the frequency distribution of the nonprobability sample the same as that of the population. Quota sampling is an example of this.
For example, the age × race distribution of the sample might be controlled to be the
same as that in the population. If we start with a large panel of volunteers, a stratified
subsample might be selected to achieve this kind of distributional balance. Each person
would receive the same weight, which is the same way that a proportionally allocated
probability sample would be treated. Considered in this way, quota sampling falls into
the quasirandomization framework.

**Example 6.2:  Assigning weights using nearest-neighbor matching.** One way of
assigning weights to a nonprobability sample is to use nearest-neighbor matching to find
a case in a probability reference survey that is similar to each nonprobability case. The
survey weight for the reference sample case is then assigned to the nonprobability case.
The `teffects` command is designed to estimate treatment effects in an observational
study where some cases are treated and others are not and is one way of doing the
matching. The "treated" are the nonprobability cases. A reference probability sample,
`ref.nhis.dta`, with 18,959 cases and a nonprobability sample, `np.nhis.dta`, with 200
cases are loaded in the code below (see `ex.6.2_nnmatch.nhis.do`). The sum of the
survey weights, `svywt`, is 58,842,608. A variable, `in_np`, is created that is 1 if the case is
in the nonprobability sample and 0 if not. The `mat_1` variable is a record counter used
to locate each reference case matched to a nonprobability case. The `ref.merge.dta`
file contains the reference cases but has only two fields: `mat_1` and the survey weight,
`svywt`.

We then reload the `ref.np.combined.dta` file, which contains both the reference
and nonprobability cases. `teffects nnmatch` finds the nearest-neighbor matches. The
full command is

```
. teffects nnmatch (y sex age_grp hisp educ race) (in_np), nneighbor(1) ///
     generate(mat_)
```

The first set of parentheses lists in order the treatment measure (`y` in this example)
and the covariates, `sex age_grp hisp educ race`, used to do the matching. The field `y`
is a useless artificial variable—we just want the matches, not an estimate of a treatment
effect. The second set of parentheses, `(in_np)`, specifies the indicator of whether a
case is a treatment or control. `nneighbor(1)` specifies that at least one match should
be made for each case. (The algorithm may actually produce many matches for each
case.) The counter numbers of the matches are stored in a series of fields prefixed with
`mat_`. The distance measure used by default is Mahalanobis, but `teffects` also offers
other options such as Euclidean that can be specified with `metric()`. Mahalanobis has
the advantage of adapting to handle covariates that may be correlated and measured
on different scales. We retain only the first match, `mat_1`, using the `drop mat_2 -`
`mat_1038` command. A refinement might be to select a match at random for each
nonprobability case.

```
. use http://www.stata-press.com/data/svywt/ref.nhis.dta, clear
. generate in_np = 0
. save ref1.nhis.dta, replace
. summarize svywt

. display r(sum)
58842608

. use http://www.stata-press.com/data/svywt/np.nhis.dta, clear
. generate in_np = 1
. save np1.nhis.dta, replace
. append using ref1.nhis.dta
. save ref.np.combined.dta, replace
.     * create a field to use in merging the nonprob and ref samples
. generate mat_1 = _n
.     * keep only ref sample cases with svywt and new variable mat_1 to use in
> merge later
. keep if in_np == 0
. keep mat_1 svywt
. rename svywt svywt1
. save ref.merge.dta, replace
. use ref.np.combined.dta, clear
. teffects nnmatch (y sex age_grp hisp educ race) (in_np), nneighbor(1)
> generate(mat_)
.     * keep nonprob cases
. keep if in_np == 1
.     * keep only the mat_1 match
. drop mat_2 - mat_1038

.     * merge nonprob and ref samples using mat_1 as merge variable
.     * many to 1 merge because some cases in ref may be used more than once in
> mat_1
. merge m:1 mat_1 using ref.merge.dta
(note: variable mat_1 was long, now double to accommodate using data´s values)

    Result                           # of obs.

    not matched                         18,892
        from master                          0  (_merge==1)
        from using                      18,892  (_merge==2)

    matched                                200  (_merge==3)


. keep if in_np == 1
. summarize svywt1

. display r(sum)
655301
```

Finally, the nonprobability sample is merged with the **ref.merge.dta** file. A many-to-1 merge is used (**merge m:1 mat_1 using ref.merge.dta**) because a record in the reference file may be matched to more than one case in the nonprobability file. This merged file is reduced down to the 200 cases in the nonprobability sample, each record of which now contains the survey weight from the matched reference case. The sum of the weights in the nonprobability sample is 655,301, which is much less than 58,842,608 from the reference file because the nonprobability sample has only 200 cases, whereas the reference has 18,959. Consequently, the weights assigned to the nonprobability sample must be scaled up to sum to 58,842,608 by calibrating with `svycal` to population controls or to estimates of them from the reference sample.

■

> **Warning**. If missing values are present in any of the covariates used for matching, a match will not be made. To be sure that all nonprobability cases are assigned a match, missing covariate values should be imputed.

Producers of the matched weights may wish to examine the distance values for outliers that might indicate a poor match. Though not cited in the literature to our knowledge, researchers may choose to discard a few outlier volunteer records to honor the "common support" assumption.

In some applications, the goal may be to combine a nonprobability sample with a probability sample, and then to make estimates for a population based on the combination. For example, Robbins, Ghosh-Dastidar, and Ramchand (Forthcoming) present a way of blending a probability sample of caregivers to wounded, ill, or injured military service members with a convenience sample of caregivers. This contrasts with how a probability reference sample is used, as discussed earlier. A reference sample is a means of estimating pseudo-inclusion probabilities for a nonprobability sample. Only the probability sample is used to make population estimations. The statistical methods for combining a probability and nonprobability sample are fairly elaborate and are not pursued further here.

## Model-based weighting

The general idea in model-based estimation when estimating a population total is to sum the responses for the sample cases and add to them the sum of predictions for nonsample cases. The key to forming unbiased estimates is that the sample and nonsample cases follow a common model that can be discovered by analyzing the sample responses. An appropriate model usually includes covariates that are known for each individual volunteer case. The covariates may or may not be known for nonsample cases, but, at a minimum, population totals of the covariates are required to construct the estimator. Suppose that a linear model for a variable $y$ is

$$E_M(y_i) = \mathbf{x}_i^T \beta$$

where the subscript $M$ means that the expectation is with respect to the model, $\mathbf{x}_i$ is a vector of $p$ covariates for unit $i$, and $\beta$ is a parameter vector. Given a sample $s$, the estimator of the slope parameter is $\widehat{\beta} = \mathbf{A}_s^{-1}\mathbf{X}_s^T\mathbf{y}_s$, where $\mathbf{A}_s = \mathbf{X}_s^T\mathbf{X}_s$, $\mathbf{X}_s$ is the $n \times p$ matrix of covariates for the sample units, and $\mathbf{y}_s$ is the $n$-vector of sample $y$'s. (If $\text{var}_M(\mathbf{y}) = \mathbf{V}$, a diagonal or nondiagonal covariance matrix, generalized least squares can be used to estimate $\beta$.) A prediction of the value of a unit in the set of nonsample units, denoted by $\bar{s}$, is $\widehat{y}_i = \mathbf{x}_i^T\widehat{\beta}$. A predictor of the population total, $t_y$, is

$$\widehat{t}_{y1} = \sum_{i \in s} y_i + \sum_{i \in \bar{s}} \widehat{y}_i \tag{6.2}$$

The theory for this prediction approach is extensively covered in Valliant, Dorfman, and Royall (2000). As written above, we would need the values of the covariates for each individual nonsample unit to construct the estimator, but the second term can also be written as $\sum_{\bar{s}} \widehat{y}_i = \mathbf{t}_{\bar{s}x}^T\widehat{\beta}$, where $\mathbf{t}_{\bar{s}x}$ is the nonsample total of the $\mathbf{x}$'s. If the population total, $\mathbf{t}_{Ux}$, is known from a census or some external data source, the nonsample total can be found by subtracting the sample totals from the population total, that is, $\mathbf{t}_{\bar{s}x} = \mathbf{t}_{Ux} - \mathbf{t}_{sx}$. If the sample is a small fraction of the population, as would be the case for most opt-in web surveys, the prediction estimator is approximately the same as predicting the value for every unit in the population and summing the predictions:

$$\widehat{t}_{y2} = \sum_{i \in U} \widehat{y}_i = \mathbf{t}_{Ux}^T\widehat{\beta} \tag{6.3}$$

The estimators in (6.2) or (6.3) are quite flexible in what covariates can be included. For example, we might predict the amount that people have saved for retirement based on their occupation, years of education, marital status, age, number of children they have, and region of the country in which they live. Constructing the estimator would require that census counts be available for each of those covariates. Another possibility is to use estimates from a larger or more accurate survey (for example, see Dever [2008] and Dever and Valliant [2010, 2016]). The reference surveys mentioned earlier could be a source of estimated control totals in which webographic covariates might be used.

Both (6.2) and (6.3) can be written so that they are weighted sums of $y$'s. If (6.2) is used, the weight for unit $i$ is

$$w_{1i} = 1 + \mathbf{t}_{\bar{s}x}^T\mathbf{A}_s^{-1}\mathbf{x}_i \tag{6.4}$$

In (6.3), the weight is

$$w_{2i} = \mathbf{t}_{Ux}^T\mathbf{A}_s^{-1}\mathbf{x}_i$$

The estimated total for an analysis variable can be written as $\widehat{t}_y = \sum_s w_i y_i$, where $w_i$ is either $w_{1i}$ or $w_{2i}$. Notice that these weights depend only on the $x$'s, not on $y$. As a result, the same set of weights could be used for all estimates. It is true that a single

set of weights will not be equally efficient for every $y$, but this situation is also true
for the design-based weights discussed in other chapters. A mean or proportion can be
estimated using the standard approach of dividing an estimate of a total by the sum of
the weights

$$\widehat{\bar{y}} = \widehat{t}_y / \widehat{N}$$

where $\widehat{N} = \sum_s w_i$ and $w_i$ is again either of the model-based weights defined above. We
denote the denominator as $\widehat{N}$ because the sum of these weights will be near $N$ in all
situations as long as an intercept is included among the model covariates. The reason
for this is somewhat technical and will only be sketched here. For the weights in (6.4),
the sum is $\sum_s w_{1i} = n + (N - n)\overline{\mathbf{x}}_{\overline{s}}^T \widehat{\beta}_x$, where $\overline{\mathbf{x}}_{\overline{s}}$ is the mean per unit of the vector of
$x$'s in the nonsample and $\widehat{\beta}_x = \mathbf{A}_s^{-1}\mathbf{t}_{sx}$ with $\mathbf{t}_{sx}$ being the sum of the $x$'s for the sample.
The term $\widehat{\beta}_x$ can be interpreted as the slope in a regression model, where a vector of 1s
is the dependent variable and the independent variables are the $x$'s plus an intercept.
Because the dependent variable is a constant, the regression coefficients on the $x$'s will
estimate 0 while the intercept will be 1. Thus, $\widehat{\beta}_x \doteq (1, 0, 0, \ldots, 0)$. The mean in the
first component of $\overline{\mathbf{x}}_{\overline{s}}$ is 1 because it corresponds to the intercept. As a result, $\overline{\mathbf{x}}_{\overline{s}}^T \widehat{\beta}_x \doteq 1$
and $\sum_s w_{1i} \doteq n + (N - n) = N$. For similar reasons, $\sum_s w_{2i} \doteq N$.

**Example 6.3: Superpopulation weights for a nonprobability survey.** The nonprobability sample in this example, `np.nhis.dta`, is the same one as in example 6.2. The
full set of code is in `ex.6.3_superpop.wts.do`. The weights in (6.4) can be computed using `svycal` by setting all weights in the nonprobability dataset to 1 in the line `generate`
`_one = 1`. The control totals in thousands were tabulated from `ref.nhis.dta`, which
was also used in example 6.2. `svycal` is then called with `_one` as the `pw` weight field.
The calibration covariates are the same ones used in the matching example. We use
`#delimit ;` as a way to break up the long `totals(...)` option into separate lines
for each covariate. Notice that a total for every level of each covariate is included,
even though `svycal` will internally drop the first level of each to solve the calibration
equations.

```
. use http://www.stata-press.com/data/svywt/np.nhis.dta, clear
. generate _one = 1
.        * control totals are divided by 1000 compared with weighted
.        * sums in ref.nhis.dta
. #delimit ;
delimiter now ;
. svycal regress i.sex i.age_grp i.hisp i.educ i.race [pw=_one],
> gen(mwt) totals(_cons=58842
>                 1.sex=28625 2.sex=30217
>                 1.age_grp=11498 2.age_grp=6418 3.age_grp=18072 4.age_grp=15231
>  5.age_grp=7621
>                 1.hisp=7979 2.hisp=41096 3.hisp=7428 4.hisp=2337
>                 1.educ=33875 2.educ=13594 3.educ=7559 4.educ=3812
>                 1.race=48622 2.race=7644 3.race=2575) ;
. #delimit cr
delimiter now cr
```

```
. summarize mwt
    Variable |        Obs        Mean    Std. Dev.        Min         Max
-------------+--------------------------------------------------------------
         mwt |        200      294.21     83.77073    110.8476    416.6757
. display r(sum)
58842
```

The sum of the calibrated weights is 58,842, which is the sum of the controls, as verified in the last command. Note that the weights are properly scaled for estimating population totals—unlike the ones from the matching example that have to be further calibrated.                                                                                  ∎

# 6.3   Variance estimation for nonprobability surveys

## Quasirandomization-based estimators

In quasirandomization weighting, we computed pseudoprobabilities of being observed in the sample, implying that the same variance formulas as in design-based weighting can be used. The distribution for inference is the one generated as if the sample units were observed repeatedly according to the pseudoprobability distribution. The pseudoprobabilities are not controlled by the survey designer but can be interpreted in the same way that selection probabilities were in earlier chapters. Decisions do need to be made about whether the sample should be treated as stratified or clustered—whether either of these features is appropriate will depend on the particular survey.

**Example 6.4: Using pseudoweights in SE calculations.** This example uses the file saved in example 6.1 and specifies the design in such a way that Stata treats it as a varying probability, with-replacement sample. The code is in ex.6.4_brf.internet.do. The computed SE uses the design-based formula appropriate to that design. SMOKE100 is an indicator of whether a person has smoked 100 cigarettes or more in her or his lifetime; PHYSACT is an indicator for whether a person participated in activities such as running, calisthenics, golf, gardening, or walking for exercise in the last month.

```
. use http://www.stata-press.com/data/svywt/brf.internet1.wtd.dta, clear
. label define physlabel 1 "Yes" 2 "No", replace
. label values PHYSACT physlabel
. svyset [pweight=pseudowt]
```

```
. svy: tabulate SMOKE100 PHYSACT, row se stubwidth(15)
```
(*output omitted*)

|                 |        | PHYSACT |       |
|-----------------|--------|---------|-------|
| SMOKE100        | Yes    | No      | Total |
| 1               | .7521  | .2479   | 1     |
|                 | (.065) | (.065)  |       |
| 2               | .904   | .096    | 1     |
|                 | (.032) | (.032)  |       |
| Total           | .8189  | .1811   | 1     |
|                 | (.0411)| (.0411) |       |

```
Key:   row proportion
         (linearized standard error of row proportion)
Pearson:
   Uncorrected    chi2(1)          =      7.6619
   Design-based   F(1, 199)        =      5.3315     P = 0.0220
```

∎

In the with-replacement variance estimator, the pseudoweights are treated as constant even though, in reality, they are also estimates. Consequently, a refinement would be to account for that variation when estimating SEs. Creating a series of replicate pseudoweights would be one option for doing this.

## Variance estimation for model-based estimators

There are several choices for variance estimators when using model-based weighting. These are described in Valliant, Dorfman, and Royall (2000, chap. 5 and 9). To fully define the model, we need to add a variance specification. The ones we summarize here are appropriate for models with mutually independent units. Although model-based estimators have been extended to cases where units are correlated within clusters (Valliant, Dorfman, and Royall 2000, chap. 9), these clustered structures are unnecessary for the web surveys and similar cases that we cover here. Suppose that the full model is

$$E_M(y_i) = \mathbf{x}_i^T \beta$$
$$V_M(y_i) = v_i$$

(6.5)

where $v_i$ is a variance parameter that does not have to be specifically defined. The variance estimators below will work regardless of the form of $v_i$ (as long as it is finite).

For use below, define $a_i$ to be $w_i - 1$, where $w_i$ is either $w_{1i}$ or $w_{2i}$. The variance estimators below then apply for either of the $w_{1i}$ or $w_{2i}$ weights. The prediction variance of an estimator of a total, $\widehat{t}_y$, is defined as

$$V_M\left(\widehat{t}_y - t_y\right) = \sum_{i \in s} a_i^2 v_i + \sum_{i \in \overline{s}} v_i \tag{6.6}$$

The population total of $y$, $t_y$, is subtracted on the left-hand side because the sum is random under the model. If the fraction of the sampled population is small, the second term above is inconsequential compared with the first. The variance estimators are built from the model residuals, $r_i = y_i - \mathbf{x}_i^T \widehat{\beta}$. An estimator of the dominant first term is

$$\sum_s a_i^2 \widehat{v}_i \tag{6.7}$$

where $\widehat{v}_i$ can be any of three choices

$$\widehat{v}_i = \begin{cases} r_i^2 \\ r_i^2/(1 - h_{ii}) \\ \{r_i/(1 - h_{ii})\}^2 \end{cases}$$

and $h_{ii}$ is the leverage for unit $i$, defined as the diagonal element of the hat matrix $\mathbf{H} = \mathbf{X}_s^T \mathbf{A}_s^{-1} \mathbf{X}_s$. In large samples where no $x$ is extreme, each leverage will be near zero.

To construct an estimator of the second term in (6.6), some assumption must be made about the form of the variance parameter, $v_i$. If, for example, $v_i = \sigma^2$, then $E_M\left(r_i^2\right) \doteq \sigma^2$. In that case, $\widehat{\sigma}^2 = \sum_s r_i^2/n$ estimates $\sigma^2$. Thus, an estimator of the second term in (6.6) would be $(N - n)\widehat{\sigma}^2$. Combining this estimator with the more general one in (6.7) gives this estimator of the prediction variance of $\widehat{t}_y$:

$$v\left(\widehat{t}_y\right) = \sum_s a_i^2 \widehat{v}_i + (N - n)\widehat{\sigma}^2 \tag{6.8}$$

The estimators of the first term are robust in that they are approximately model-unbiased regardless of the form of $v_i$ (which is unknown) if the sampling fraction is small. The first choice, $\widehat{v}_i = r_i^2$, when used in (6.7) and (6.8), gives an example of a sandwich estimator, which is available in most of Stata's estimation commands by using the vce(robust) option. The second choice adjusts for the fact that $r_i^2$ is slightly biased for $v_i$. The third choice is similar to the jackknife, in which one sample unit at a time is deleted, a new estimate of the total computed, and the variance among those delete-one estimates is used. Both the second and third choices were studied in Long and Ervin (2000), among many others, for regression estimation and were studied in Valliant, Dorfman, and Royall (2000) for finite population estimation. Because the second term in (6.8) is usually negligible compared with the first, misspecifying its form is likely to be unimportant.

If the population totals for some of the covariates are estimated from an independent survey, then the variance in (6.8) should be modified by adding a term to reflect that additional uncertainty (see section 4.4 and Dever and Valliant [2010, 2016]).

**Example 6.5: Using the model-based weights.** We can use the weights computed in example 6.3 to estimate the total number of persons who delayed seeking medical care in the previous year because of cost. The full code is in `ex.6.5_use.superpop.wts.do`. The `regress()` option of `svyset` is used in the same way as it was in chapter 4 for calibrated estimators from a probability sample. The population totals must be specified in the `regress()` option as they were in `svycal` in example 6.3. The SEs computed by `svy: tabulate` are those in (6.7) with $\hat{v}_i = r_i^2$, that is, only the first term of (6.8).

```
. #delimit ;
. svyset ID [pweight = mwt], regress(i.sex i.age_grp i.hisp i.educ i.race,
> totals(_cons=58842
>         1.sex=28625 2.sex=30217
>         1.age_grp=11498 2.age_grp=6418 3.age_grp=18072 4.age_grp=15231
> 5.age_grp=7621
>         1.hisp=7979 2.hisp=41096 3.hisp=7428 4.hisp=2337
>         1.educ=33875 2.educ=13594 3.educ=7559 4.educ=3812
>         1.race=48622 2.race=7644 3.race=2575));
. #delimit cr
. svy: tabulate delay_med, count se format(%12.0f)
(running tabulate on estimation sample)
```

| | | | | |
|---|---|---|---|---|
| Number of strata | = | 1 | Number of obs = | 200 |
| Number of PSUs | = | 200 | Population size = | 58,842 |
| Calibration | : | regress | Design df = | 199 |

| delay.med | count | se |
|---|---|---|
| 1 | 5309 | 1163 |
| 2 | 53533 | 1163 |
| Total | 58842 | |

```
Key:  count  =  weighted count
        se   =  linearized standard error of weighted count
```

Stata uses the same variance estimator that it would for an unstratified, varying probability sample selected with replacement that has been calibrated to the covariates.

For a mean estimated as $\widehat{\overline{y}} = \widehat{t}_y/\widehat{N}$, the denominator $\widehat{N}$ can be treated as a constant if an intercept is in the model so that $\widehat{N} \doteq N$. In that case,

$$v\left(\widehat{\overline{y}}\right) = v\left(\widehat{t}_y\right)/\widehat{N}^2$$

Notice that $\widehat{\overline{y}}$ is not treated as nonlinear as it was for design-based variance estimation in chapter 5. This is essentially what Stata does in the following code for estimating the proportion of persons who delayed medical care. Computing the SE of the proportion who delayed care (`delay.med=1`) by hand using the example above gives $\sqrt{1163^2/58842^2} = 0.019765$—almost the same as 0.019769 below.

```
. svy: tabulate delay_med, se format(%12.6f)
(running tabulate on estimation sample)

Number of strata    =          1        Number of obs    =        200
Number of PSUs      =        200        Population size  =     58,842
Calibration         :    regress        Design df        =        199


delay.med | proportion           se
----------+-------------------------------
        1 |   0.090231     0.019769
        2 |   0.909769     0.019769
----------+-------------------------------
    Total |   1.000000

  Key:  proportion  =  cell proportion
        se          =  linearized standard error of cell proportion
```

## 6.4   Bayesian approaches

Bayesian inference for finite populations is similar to the model-based approach discussed above but with an extra layer of assumptions. Parameters such as $\beta$ and $v_i$ in (6.5) are treated as having their own probabilistic distribution. This approach can lead to some interesting estimators that adapt themselves to the patterns found in the particular dataset in hand. Ghosh and Meeden (1997) and Ghosh (2009) summarize many of the results. A particular application of this approach is called hierarchical regression modeling or multilevel regression and poststratification (MRP), in which $y$ is represented as a type of poststratification model,

$$E_M\left(y_i\right) = \mu_\gamma, \quad i \in U_\gamma$$

where $U_\gamma$ is the population of units in poststratum $\gamma$. The model is hierarchical in that $y$ is modeled and the parameters in the $y$ model are also modeled with prior distributions.

The poststrata can be extremely elaborate. For example, in forecasting the outcome of the U.S. 2012 presidential election, Wang et al. (2015) used all possible combinations of sex (2 categories), race (4 categories), age (4 categories), education (4 categories), state (51 categories), party ID (3 categories), ideology (3 categories), and how a person voted in the 2008 presidential election (3 categories) to create 176,256 cells. They then used Bayesian modeling to predict the mean in each cell, which was the proportion of voters who would vote for a particular candidate. The advantage of the Bayesian multilevel model is that the estimates for relatively sparse or empty cells can be improved through "borrowing strength" from demographically similar cells that have more cases. The estimated proportion of votes for a candidate was then

$$\widehat{\overline{y}} = \sum_{\gamma} P_{\gamma} \widehat{\mu}_{\gamma}$$

where $\widehat{\mu}_{\gamma}$ is the estimated proportion in the poststratum who will vote for the candidate and $P_{\gamma}$ is the proportion of voters in the cell. ($\widehat{\mu}_{\gamma}$ is the mean of a Bayesian posterior distribution.)

In cases where $P_{\gamma}$ must also be estimated, those proportions might be also modeled. Dong, Elliott, and Raghunathan (2014) and Zhou, Elliott, and Raghunathan (2016c,b,a) all address this additional complication. The methods used are exceedingly difficult to implement, and we will not illustrate them here.

## 6.5   Some general comments

There is not universal agreement that the methods we have outlined in this chapter for nonprobability samples will produce reliable estimates. Besides the AAPOR panel reports cited earlier that enumerate problems with these samples, AAPOR has also issued some comments on calculating margins of error (MOEs, essentially confidence intervals) for estimates from nonprobability samples.[14] The standard repeated-sampling arguments that justify confidence intervals do not apply to nonprobability samples because the "mechanism" for obtaining the sample is not random selection that a sampler controls. An MOE computed with a variance estimate from section 6.3 relies on a model for either the probability of being in the sample or for an analysis variable $y$ to be correct. AAPOR cautions that the assumptions in these models may be wrong and, as a result, so are the MOEs.

Of course, in probability samples with high nonresponse, the repeated-sampling assumptions do not hold either. Thus, making inferences can boil down to picking the approach with assumptions you feel are the least violated.

If Bayesian methods are used, the interval estimates are called "credible intervals"; these provide another attempt to convey the statistical uncertainty in an estimate.[15] A credible interval is not interpreted in the same way as an MOE but, similar to an MOE, is a range where a population value is likely to fall. As for a model-based confidence interval, the accuracy of a credible interval depends on how well the model(s) that generated it applies to the population as a whole. Models that fit very well for a particular sample may not predict well for the nonsample part of the population.

---

14. See http://www.aapor.org/Education-Resources/Election-Polling-Resources/Margin-of-Sampling-Error-Credibility-Interval.aspx.

15. See http://www.aapor.org/Publications-Media/Press-Releases/Archived-Press-Releases/Understanding-a-credibility-interval%E2%80%9D-and-how-it-d.aspx.

# 7 Weighting for some special cases

This chapter touches on a few special situations in survey weighting. The first (section 7.1) is using normalized weights, which are scaled to sum to the unweighted sample size. In section 7.2, we discuss cases where a dataset is supplied with multiple weights for different purposes. Section 7.3 covers two-phase sampling, which is often used when expending extra effort to convince nonrespondents to participate in a survey. Some surveys use more than one frame for sampling; section 7.4 discusses composite weights for combining the estimates from the different frames. Section 7.5 briefly mentions the effects on analyses of masking strata and PSU IDs. Finally, in section 7.6, we give an extended discussion with examples of an important issue for data analysis—whether to use survey weights when fitting regression models.

## 7.1 Normalized weights

In this book, we consider weights that are appropriately scaled for estimating population totals. Thus, a weighted sum of data, $\hat{t}_y = \sum_{i \in s} w_i y_i$, will estimate the population total, $t_U = \sum_{i \in U} y_i$. The sum of the weights themselves would be an estimate of the population size, $N$, that is, $\hat{N} = \sum_{i \in s} w_i$. However, it is not unusual for analysts to normalize weights to sum to the sample size. That is, $\sum_{i \in s} w_i^* = n$, where $n$ is the sample size of units (typically people) used for analysis and $w_i^* = a_i^* w_i$, with $a_i^*$ being the scaling factor. In particular, $a_i^* = n / \sum w_i$.

To some extent, normalizing weights is an anachronism from days when software packages had no special procedures for analyzing survey data. For example, if nonsurvey software were used to fit a linear regression model, the degrees of freedom for the error sum of squares may be reported as the sum of the weights minus the number of parameters. In the United States, the population size in 2017 was over 320 million, which would also be the reported degrees of freedom in a model fit from a population census, assuming the sum of the weights estimates $N$. This essentially infinite degrees of freedom would be reported whether the number of sample persons was 100, 1,000, or 10,000 and regardless of the type of sample design. Forcing the weights to sum to $n$ avoids this sort of misleading output.

However, a number of good packages (like Stata) are now available that have procedures for correctly analyzing complex survey data. These packages all avoid the degree-of-freedom gaffe mentioned above. Additionally, they will correctly estimate descriptive statistics, model parameters, and their standard errors when the regular (nonnormalized) survey weights are used. There are several reasons for not normalizing weights. The first is that population totals cannot be estimated using normalized weights. The second is that the ability to perform some quality checks on the weights by comparing their sums with external totals, like population census counts, is lost. The third is that having normalized weights may induce some analysts to use inappropriate nonsurvey procedures, possibly with some adjustments, to analyze survey data. This approach is sometimes suggested in the education literature (for example, Hahs-Vaughn [2005]). This last objection is, perhaps, less persuasive because analysts have become more skilled at using survey data.

One case in which normalized weights have a computational advantage is in the estimation of hierarchical linear models (Pfeffermann et al. 1998). However, some hierarchical linear model procedures such as `mixed` in Stata or the `hlm` package (SSI 2015) will scale the weights for you. As a result, there is no need to normalize the weights in advance.

## 7.2   Multiple weights

The techniques discussed in earlier chapters are oriented toward one multipurpose analysis weight used for all analyses. However, in some applications, more than one weight may be provided for each record for special purposes. We provide a few examples below:

- The National Health and Nutrition Examination Survey (NHANES; the source data used in several examples throughout this text) is an on-going cross-sectional survey conducted in the United States since the early 1960s to "assess the health and nutritional status of adults and children" through data collected via interviews and physical exams.[1] Participants are randomly selected in four stages: U.S. county or county group, geographic segment within PSU, household within segment, and one to two persons per household. NHANES analyses may be conducted with one of nine survey weights. The NHANES documentation provides detailed guidance the associated analyses with each; this includes, for example, two weights devoted to the study of allergies and the central nervous system with data collected on a random subsample of respondents.

- The European Social Survey (ESS) is a cross-national survey conducted every two years across Europe.[2] The purpose of the ESS is to collect attitude and behavior patterns on a variety of subjects, including immigration, climate change and politics. The sample design varies by country depending on the presence of a high-quality population registry. The ESS documentation provides details on weights used for country-specific analyses; a separate set of weights is used for data combined across two or more countries.

- The High School Longitudinal Survey of 2009 (HSLS:09) is a nationally representative longitudinal study of U.S. students to measure algebraic and problem solving skills. Students in the ninth grade are randomly chosen from participating

---

1. See https://www.cdc.gov/nchs/nhanes/.
2. See http://www.europeansocialsurvey.org/.

public, private, and charter schools selected in 2009 (base year).[3] Two follow-up interviews were conducted with the students to evaluate their educational and workforce trajectory. The HSLS:09 documentation contains details on five weights for analysis with the base-year interviews: school director, student (plus algebraic assessment), parent, science teacher, and mathematics teacher. To date, weights are also available to conduct longitudinal analyses from base year to first follow-up with students and parents.

- The Early Childhood Longitudinal Study, Kindergarten Class of 1998–1999 (ECLS-K) is a nationally representative longitudinal study of students in U.S. kindergartens to understand "children's development, early learning, and performance in school".[4] Kindergartners were randomly chosen in 1998 (base year) from participating schools selected from geographic PSUs. Base-year respondents were contacted four times (first, third, fifth, and eighth grades) for subsequent interviews. The ECLS-K February 2009 documentation details five cross-sectional weights all linked to student-level analyses with various combinations of contextual data. For example, only a random subsample of students has interview data from mathematics and science teachers. Additionally, there are 12 longitudinal weights for response combinations across the cycles (for example, child response in kindergarten, third, fifth, and eighth grades only, but not first grade).

The decision to produce multiple weights should follow from the sampling and analysis plans and address the needs of data users. There are a few important questions to answer before proceeding with the calculations.

*Which sample units are randomly included or excluded in the construction of the special-purpose weight?* Because of exclusions, we introduce a fifth group to the AAPOR response categories—not applicable (NA). This group includes, for example, units eligible but not selected for a special study such as those randomly excluded from the NHANES allergy evaluation. This is known as a two-phase sampling design (see section 7.3). Additionally, the AAPOR response categories should be revisited to determine, for example, those original sample units not eligible for the special study. For example, interview responses or demographic and physiological characteristics may preclude participation; for example, women confirmed or suspected of being pregnant could be excluded from specialized medical examinations.

*What is the definition of nonresponse for longitudinal studies?* Longitudinal studies, like ECLS-K, follow a cohort of sample units over a specified period (see, for example, Lynn [2009]). Many of these studies include sample members even if they do not respond to each interview cycle. With ECLS-K, for example, the sample members were required to participate in the base year but were retained for analyses even if they missed an interview. The analysis plan should include a description of the intended analyses across multiple years with an associated definition of response. Finally, nonresponse models used for the main study should be revisited to determine any changes.

---

3. See https://nces.ed.gov/surveys/hsls09/.
4. See https://nces.ed.gov/ecls/kindergarten.asp.

*Are different population control totals needed?* The key to answering this question is related to the unit of analysis. With HSLS:09, for example, one of the five survey weights is used only for school-level analysis involving the school administrator and counselor interviews. The remaining four are for student-level analyses alone or in combination with contextual information obtained from the parent or teacher interviews.

As emphasized in chapter 8, study documentation is critical for appropriate use of the weights for specialized analyses.

## 7.3   Two-phase sampling

Two-phase (or double) sampling occurs when a subsample is selected from the original sample based on information obtained during an initial data collection (see, for example, Fuller [2009, sec. 3.3]). For example, the NHANES design includes four random subsamples for certain physical measurements based on information from the household interview.

Another type of two-phase sampling is sampling for nonresponse or nonresponse follow-up (NRFU). Eligible phase 1 nonrespondents form the phase 2 sampling frame; a subsample is then selected within the phase 1 strata that may include additional stratification based on phase 1 information. For example, paradata (Kreuter, Couper, and Lyberg 2010) may be used to oversample households indicating the presence of children for a study involving childhood vaccinations and in-home injuries.

Second-phase sampling weights are calculated just like base weights, which were discussed in section 2.1. For example, if a simple random sample is selected within the phase 1 strata only among the phase 1 nonrespondents, then the conditional phase 2 base weight in stratum $h$ is

$$
a_i^{(2)} = \begin{cases}
1 & \text{if unit } i \text{ is a phase 1 respondent} \\
0 & \text{if unit } i \text{ is not eligible for phase 2} \\
  & \text{or is not sampled} \\
N_{s_{\text{ENR}},h}^{(2)}/n_{s_{\text{ENR}},h}^{(2)} & \text{if unit } i \text{ is selected for phase 2}
\end{cases}
$$

where $N_{s_{\text{ENR}},h}^{(2)}$ is the number of phase 1 NRs eligible for phase 2 in stratum $h$, and $n_{s_{\text{ENR}},h}^{(2)}$ is the number subsampled for phase 2 from that stratum. This subsampling adjustment is applied to the phase 1 weight at the point in which sampling occurred. Using an NHANES special study as an example, the subsampling adjustment may be applied to the phase 1 base weight adjusted for nonresponse or even calibration. You would use the phase 1 base weight before nonresponse adjustment for NRFU.

Variance estimation for a study with two phases of sampling can be quite complicated if otherwise simplifying assumptions are not appropriate. Interested readers are referred to Fuller (2009, sec. 3.3) and other such sources for additional details.

## 7.4   Composite weights

Composite weights are used to combine samples drawn from partially or entirely overlapping subpopulations within a common target population. For example, CHIS 2013–2014 (Flores-Cervantes, Norman, and Brick 2014) was a telephone survey conducted in the United States state of California with samples selected independently from landline telephone frames, a cellular telephone frame, and an address-based sampling frame (Iannacchione 2011). In this instance, the overlap occurs because households could have only a landline telephone, only a cellular telephone, both a landline and cellular telephone, or for a small percent, no telephone at all. Note that some units from each frame are unique to that frame (for example, landline-only households selected from a landline frame).

For a dual-frame survey where only two sampling frames are used, the composite weight for units in stratum $h$ common to both frames is constructed as:

$$
w_i^{\text{comp}} = \begin{cases} w_i & \text{if unit } i \text{ is unique to either frames A or B} \\ \lambda_h w_i & \text{if unit } i \text{ is selected from frame A} \\ & \quad \text{and in the overlap} \\ (1 - \lambda_h) w_i & \text{if unit } i \text{ is selected from frame B} \\ & \quad \text{and in the overlap} \end{cases}
$$

Several recommendations exist for the appropriate choice of the $\lambda_h$ (the composite factor) depending on the survey needs (see, for example, Lohr and Rao [2006]; Lohr and Brick [2014]; and Wolter et al. [2015]). The relative contribution from the frame 1 sample to the overall variance (for frames with complete overlap) is one example of a composite factor, but it depends on a single $y$-variable. Brick et al. (2011) provide a method for combining samples to minimize bias for a set of variables. Other researchers have simply used the relative sample size (that is, $\lambda_h = n_{Ah}/(n_{Ah} + n_{Bh})$ where $n_{Ah}$ and $n_{Bh}$ are the sample size from frames A and B in common stratum $h$, respectively) as a surrogate (see, for example, Flores-Cervantes, Norman, and Brick [2014]). Calibration may be applied to the input weights for this process or after the composite weights are constructed.

## 7.5   Masked strata and PSU IDs

Masking means that actual strata and PSU IDs are not disclosed. Masking to protect identities of respondents is especially common in public-use files supplied by governmental agencies. Although strata and PSU IDs may be supplied for SE calculations, the intention is that they cannot be connected to the real units. Another method of confidentiality protection is to suppress all strata and PSU IDs and only release replicate weights. Section 5.4 details three replication variance techniques—jackknife, balanced repeated replication (BRR), and bootstrap. In addition to capturing the variability associated with the random weight adjustments, replication methods also allow the masking of the original design components as one method to protect the identities of the participating sample members. Example 5.4 demonstrates this very trait—only the full-sample weight and the jackknife weights were used in the analysis.

Data files with many strata and PSUs, however, will generate a correspondingly large number of replicate weights. The solution is to group PSUs (VarUnit or variance unit) and possibly strata (VarStrat or variance stratum) to produce a more reasonable number of replicates. However, as cautioned in section 5.4.4, collapsing should be implemented in such as a way as to maintain approximate unbiasedness and consistency. For example, strata or PSUs should be combined based only on design variables—not by looking at the $y$ variables. Combining units that are similar with respect to values of key analysis variables will tend to bias SE estimates downward (Rust 1985; Rust and Kalton 1987). See the appendices in Westat (2007) for additional information. When using files with replicate weights constructed by someone else, an analyst generally has no choice but to trust that the database constructor has properly combined such units.

Some loss of information usually accompanies masking. Geographic location of a sample element and the size of the PSU containing each element may be lost unless separate regions or similar fields are in the dataset. For example, whether an element is in a certainty or noncertainty PSU is likely to be masked. In some applications of replication variance estimation, no strata or PSUs IDs may be supplied at all, as noted above. This was the case in example 5.9, where bootstrap weights were provided for the National Maternal and Infant Health Study (NMIHS). Although the replication weights permit correct estimation of SEs for most or all statistics, having only replication weights may limit the options for accounting for parts of sample designs when fitting models, as described in section 7.6.

## 7.6   Use of weights in fitting models

Whether to use survey weights when fitting models is a perennial source of debate. The basis of dispute is that the goals in model fitting are different from those in estimating finite population descriptive parameters, like means or totals. When estimating a mean, total, quantile, or other descriptive statistic, the methods in previous chapters apply. Survey weights are used to inflate a sample to a population. Design-based theory generally guides the weight construction, although there is a separate literature on model-based estimation of descriptive statistics (for example, see Ghosh [2009] and Valliant, Dorfman, and Royall [2000]).

When fitting models, the goal is usually to describe a structure that holds more broadly than just for a finite population at a particular point in time. For example, suppose that an analysis is geared toward finding out how well a linear model fits the population

$$y_i = \mathbf{x}_i^T \beta + \varepsilon_i, i \in U$$

where $y_i$ is an analysis variable for unit $i$, $\mathbf{x}_i$ is a vector of $p$ covariates for unit $i$, $\beta$ is the vector parameter to be estimated, and $\varepsilon_i$ is a random error with mean 0. There are different ways of estimating $\beta$ that are reviewed in Binder and Roberts (2009) and Pfeffermann et al. (1998), among others. One method is the pseudolikelihood approach, in which the "census" estimating equations should be the target of estimation. That is, we should estimate the slope that would be found in ordinary least-squares (OLS) estimation (assuming that all $\varepsilon_i$ have the same model variance) if a census of the finite population were to be done. This leads to the set of census estimating equations

$$\sum_{i \in U} \mathbf{x}_i \left( y_i - \mathbf{x}_i^T \beta \right) = 0 \tag{7.1}$$

Equation (7.1) can be solved explicitly as

$$\widehat{\beta} = \left( \mathbf{X}_U^T \mathbf{X}_U \right)^{-1} \mathbf{X}_U^T \mathbf{y}_U$$

where $\mathbf{X}_U$ is the $N \times p$ matrix of covariates for the entire finite population and $\mathbf{y}_U$ is the $N$-vector of $y$'s for all units in the population. Because (7.1) is a population total, it can be estimated with

$$\sum_{i \in s} w_i \mathbf{x}_i \left( y_i - \mathbf{x}_i^T \beta \right) \tag{7.2}$$

assuming that the $w_i$ are scaled for estimating totals (that is, not normalized as in section 7.1). Setting (7.2) to 0 and solving gives the survey-weighted least-squares estimator

$$\widehat{\beta}_w = \left( \mathbf{X}^T \mathbf{W} \mathbf{X} \right)^{-1} \mathbf{X}^T \mathbf{W} \mathbf{y}$$

where $\mathbf{X}$ is the matrix of covariates for the sample, $\mathbf{W}$ is the $n \times n$ diagonal matrix of weights for the sample units, and $\mathbf{y}$ is the sample vector of analysis variables. This estimator is approximately unbiased and consistent for the census parameter with respect to repeated sampling under whatever design was used. If the weights involve nonresponse adjustments or calibration, as discussed in chapter 4, the properties are over both repeated random sampling and repeated random responding. If a nonprobability sample is selected, as in chapter 6, then the properties are strictly quasirandomization or superpopulation model based.

If a logistic, probit, or some other nonlinear model is fit, a set of estimating equations like those in (7.2) can be formed. An explicit solution for a parameter estimate cannot be found, but the estimating equations can be solved iteratively.

Even if the model is not specified correctly, the solution to (7.2) and similar survey-weighted estimating equations can be interpreted as an estimate of the model that would be fit if a census were done. Some writers such as Kish and Frankel (1974) see this clear interpretation as an advantage. Others would maintain that, unless the model is correctly specified, the parameter estimates are meaningless. There are arguments for not using survey weights in modeling if important survey design features are accounted for in other ways. Pfeffermann (1993) reviews these issues. Gelman (2007) and Breidt and Opsomer (2007) are more recent sources. Besides use of weights, an analyst must consider whether design features like strata and clusters need to be accounted for when modeling.

The key technical concept is ignorability of the sample design—an idea defined by Rubin (1976). Also see Little (1982) and Valliant, Dorfman, and Royall (2000, sec. 2.6). Roughly speaking, the sample design is ignorable if the probability of being in the sample does not depend on any $y$ values. A pure probability sample, with inclusion probabilities depending only on covariates, like strata and cluster membership and measures of size, meets the criteria to be ignorable. In a nonprobability sample, there can be some doubt about ignorability, especially if units have volunteered for the sample, because their likelihood of volunteering might depend on the $y$'s measured in the survey.

If you are modeling the expected value of some $y$ by a function $m(\mathbf{x}; \beta)$, "correct specification" means getting the correct form of $m(\cdot)$ and including the right set of covariates $\mathbf{x}$. Even if the sample design is ignorable, the same model that holds for the sample may not also hold for the full population. In that case, some design features might need to be in the model. For example, if a separate model is appropriate within design strata, the strata must be interacted with other covariates. Covariates that affect sample selection or response should also be included if they are also predictive of outcome variables. If units within clusters have similarities, the clustering should be accounted for when estimating the SEs of parameter estimates.

In some cases, detailed sample design information may not be available to analysts, especially at the individual element level. This will be the case when masked strata or PSU IDs are used as in section 7.5. In those situations, including the survey weight (or the selection probability) as an independent variable has been suggested (Rubin 1983). Although including $\pi_i$ or $w_i$ as an independent variable may lead to model-unbiased predictions of $E_M(y_i)$, these covariates are typically not of scientific interest. If, for example, the analytic goal is to use survey data to see whether a smoking cessation program works better for males than females, having the survey weight or selection probability in the regression equation is more of a nuisance than a help. On the other hand, if the goal is prediction, then using $\pi_i$ or $w_i$ as a covariate is advisable if that is the only way to obtain unbiased predictions, $\widehat{y}_i$.

## 7.6.1 Comparing weighted and unweighted model fits

Korn and Graubard (1999, sec. 4.4–4.6) give some practical advice on whether unweighted estimates of model parameters are preferable to weighted ones. One measure they suggest computing is an inefficiency measure defined as

$$\text{Ineff} = 1 - \frac{v(\widehat{\beta}_{\text{OLS},j})}{v(\widehat{\beta}_{wj})}$$

where $v(\widehat{\beta}_{\mathrm{OLS},j})$ is the estimated variance of the OLS estimator of the $j$th coefficient in a regression and $v(\widehat{\beta}_{wj})$ is the estimated variance of the survey-weighted estimate. Ineff will be in $[0,1]$ if the unweighted estimate has a smaller variance than the weighted. But, it is possible for the estimated variance of the weighted estimate to be less than that of the unweighted, in which case Ineff can be negative.

An important caveat in determining if Ineff is a useful measure is whether the regression model is correctly specified. If it is not, then comparing variances of two different ways of fitting the wrong model seems pointless. On the other hand, you may think that if the model is misspecified, then the weighted estimates at least estimate the census parameter. If so, then you should use the weighted estimates. In which case, Ineff is not useful. Only when an analyst feels that the model is correctly specified is Ineff useful.

**Example 7.1: Weighted versus unweighted regression model estimates.** This example looks at the differences in SEs of estimated coefficients in a logistic regression model. The dataset used is from the 2003 National Health Interview Survey, which is provided in the **PracTools** R package. Whether a person delayed medical care in the previous 12 months is modeled as a function of age, whether a person is Hispanic, whether a person's parents live with her or him, education level, and race. When computing Ineff, a choice must be made for which design features should be incorporated into $v(\widehat{\beta}_{\mathrm{OLS},j})$. This example sets all weights to 1, but the SE for the OLS estimate accounts for strata and PSUs. The estimates, $v(\widehat{\beta}_{wj})$, account for strata, PSUs, and weights. (The code can be found in **ex.7.1_wts.in.reg.do**.)

```
. use http://www.stata-press.com/data/svywt/nhis.large.dta
. svyset psu [pw=svywt], strata(stratum)
. generate delay = abs(delay_med - 2)
. svy: logit delay i.age_grp i.hisp i.parents i.educ i.race
. predict pred_wt, pr
. generate one = 1
. svyset psu [pw=one], strata(stratum)
. svy: logit delay i.age_grp i.hisp i.parents i.educ i.race
. predict pred_unwt, pr
. twoway scatter pred_unwt pred_wt || line pred_unwt pred_unwt, legend(off)
> ytitle("Weighted predictions") xtitle("Unweighted predictions")
```

∎

Figure 7.1 shows the weighted predictions versus the unweighted. There is a noticeable difference for larger probabilities where the weighted predictions are smaller than
the unweighted. Thus, if predicted probabilities were important, the decision of whether
to use a weighted or unweighted model is more consequential.

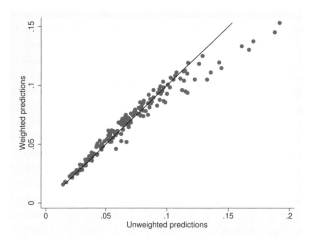

Figure 7.1. Comparison of predictions from weighted and unweighted logistic regression
for delaying medical care due to cost. Reference line at $45^o$.

Table 7.1 shows the results. The weighted and unweighted coefficients have the same
sign but in many cases have noticeably different sizes. SEs for unweighted estimates are
less than those for weighted estimates for 11 of 13 parameters. The only case for which
the significance at the 0.05 level differs is `hisp.2`. The inefficiencies range from 0.28 to
slightly negative. The loss on the SE scale from the weighted regression is largest for
`hisp.3` where it is 18%, but the coefficient for that level of Hispanic is not significant
in either the weighted or unweighted regressions. Because the increases in SEs from
weighted regression are fairly small, the advantages of using unweighted regression are
minimal.

Table 7.1. Comparison of weighted and unweighted logistic regression with an NHIS dataset

|  | Weighted estimates | | | | Unweighted estimates | | | | |
|---|---|---|---|---|---|---|---|---|---|
|  | Coef. | SE | $t$ | $P>t$ | Coef. | SE | $t$ | $P>t$ | Ineff |
| age_grp |  |  |  |  |  |  |  |  |  |
| 2 | 0.851 | 0.158 | 5.38 | 0 | 0.811 | 0.135 | 5.98 | 0 | 0.27 |
| 3 | 0.825 | 0.142 | 5.82 | 0 | 0.748 | 0.132 | 5.67 | 0 | 0.14 |
| 4 | 0.638 | 0.149 | 4.28 | 0 | 0.649 | 0.135 | 4.79 | 0 | 0.18 |
| 5 | −0.346 | 0.158 | −2.19 | 0.031 | −0.385 | 0.147 | −2.63 | 0.01 | 0.14 |
| hisp |  |  |  |  |  |  |  |  |  |
| 2 | 0.202 | 0.099 | 2.05 | 0.044 | 0.134 | 0.094 | 1.44 | 0.155 | 0.10 |
| 3 | −0.588 | 0.482 | −1.22 | 0.226 | −40.380 | 0.408 | −0.93 | 0.354 | 0.28 |
| 4 | −0.246 | 0.596 | −0.41 | 0.681 | −0.340 | 0.604 | −0.56 | 0.575 | −0.03 |
| 2.parents | 0.416 | 0.097 | 4.28 | 0 | 0.370 | 0.085 | 4.37 | 0 | 0.24 |
| educ |  |  |  |  |  |  |  |  |  |
| 2 | −0.145 | 0.084 | −1.73 | 0.088 | −0.126 | 0.078 | −1.62 | 0.11 | 0.14 |
| 3 | −0.317 | 0.100 | −3.18 | 0.002 | −0.271 | 0.102 | −2.66 | 0.01 | −0.04 |
| 4 | −0.576 | 0.172 | −3.35 | 0.001 | −0.482 | 0.160 | −3.02 | 0.003 | 0.14 |
| race |  |  |  |  |  |  |  |  |  |
| 2 | 0.676 | 0.486 | 1.39 | 0.169 | 0.370 | 0.419 | 0.88 | 0.379 | 0.26 |
| 3 | 0.095 | 0.545 | 0.17 | 0.863 | 0.079 | 0.551 | 0.14 | 0.886 | −0.02 |
| _cons | −3.380 | 0.133 | −25.47 | 0 | −3.261 | 0.115 | −28.31 | 0 | 0.25 |

## 7.6.2   Testing whether to use weights

Directly determining that a sample design (especially one that involves nonresponse) is ignorable is usually impossible. However, there are some indirect checks reviewed by Pfeffermann and Sverchkov (2009) for determining whether weighted and unweighted parameter estimates are statistically different from each other. For linear regression, one method is to fit this extended model:

$$\mathbf{y} = \gamma\mathbf{X} + \tau\mathbf{X}\mathbf{W} + \varepsilon$$

Note that the matrix $\mathbf{X}$ contains an intercept; therefore, the survey weight and the weighted covariates are included in the model through $\mathbf{X}\mathbf{W}$. The unweighted estimate $\widehat{\tau}$ is equal to $\mathbf{A}\left(\widehat{\beta}_{\text{OLS}} - \widehat{\beta}_{w}\right)$, where

$$\mathbf{A} = \{\mathbf{X}^{T}\mathbf{W}\mathbf{W}\mathbf{X} - \mathbf{X}^{T}\mathbf{W}\mathbf{X}\left(\mathbf{X}^{T}\mathbf{X}\right)^{-1}\mathbf{X}^{T}\mathbf{W}\mathbf{X}\}^{-1}\left(\mathbf{X}^{T}\mathbf{W}\mathbf{X}\right)$$

(DuMouchel and Duncan 1983; Fuller 1984; Korn and Graubard 1999). Thus, a test of $H_0 : \tau = 0$ will be a test of $H_0 : \beta_w = \beta_{\text{OLS}}$. Because the estimates of $\gamma$ and $\tau$ will be model-unbiased regardless of the covariance structure of $\varepsilon$, this model can be fit by OLS; see the following example.

**Example 7.2: Testing use of weights in a linear regression.** We use the 2009 Residential Energy Consumption Survey (RECS) dataset from example 5.8. The total British thermal units (BTUs) in millions consumed by a household for cooling (`totbtucolM`) is modeled as a function of cooling degree days (number of days where the temperature is above 65°F (CCD65), climate zone (`AIA_Zone`), total number of rooms in the housing unit (`TOTROOMS`), type of cooling system (`rCOOLTYPE`), and total square feet in the housing unit (`TOTSQFT`). The `w_*` variables are the independent variables in the model times the final weight, `NWEIGHT`. (The code is in `ex.7.2_svywt.test.linear.do`.) First, a survey-weighted regression is run using

```
svy: regress totbtucolM CDD65 i.AIA_Zone TOTROOMS i.rCOOLTYPE TOTSQFT
```

The `svy: regress` procedure does not provide an $R^2$ value, but an adjusted $R^2$ value is computed as `adjr2`.[5] The adjusted $R^2$ from the weighted regression is 0.5451. An unweighted regression is then run using

```
regress totbtucolM CDD65 i.AIA_Zone TOTROOMS i.rCOOLTYPE TOTSQFT
```

The coefficient estimates are very similar and adjusted $R^2 = 0.5474$.

```
. use http://www.stata-press.com/data/svywt/recs2009, clear
. generate one = 1
. generate totbtucolM = TOTALBTUCOL/1000
. generate w_one = NWEIGHT*one
. generate w_cdd65 = NWEIGHT*CDD65
. generate w_aiazone1 = NWEIGHT*(AIA_Zone == 1)
. generate w_aiazone2 = NWEIGHT*(AIA_Zone == 2)
. generate w_aiazone3 = NWEIGHT*(AIA_Zone == 3)
. generate w_aiazone4 = NWEIGHT*(AIA_Zone == 4)
. generate w_aiazone5 = NWEIGHT*(AIA_Zone == 5)
. generate w_totrooms = NWEIGHT*TOTROOMS
. recode COOLTYPE (-2 = 0), gen(rCOOLTYPE)
. generate w_cooltype1 = NWEIGHT*(COOLTYPE==1)
. generate w_cooltype2 = NWEIGHT*(COOLTYPE==2)
. generate w_cooltype3 = NWEIGHT*(COOLTYPE==3)
. generate w_cooltype0 = NWEIGHT*(COOLTYPE < 0)
. generate w_totsqft = NWEIGHT*TOTSQFT
.      * wtd regn
. svyset [pw=NWEIGHT], vce(brr) brrw(brr_weight_1 - brr_weight_244) fay(0.5)
> dof(244) mse
```

---

5. Based on code posted by Steven Samuels on the Statalist,
   http://www.stata.com/statalist/archive/2008-10/msg00850.html.

. svy: regress totbtucolM CDD65 i.AIA_Zone TOTROOMS i.rCOOLTYPE TOTSQFT

(*output omitted*)

| totbtucolM | Coef. | BRR *<br>Std. Err. | t | P>\|t\| | [95% Conf. Interval] | |
|---|---|---|---|---|---|---|
| CDD65 | .0028506 | .0002234 | 12.76 | 0.000 | .0024106 | .0032906 |
| AIA_Zone | | | | | | |
| 2 | .2375407 | .1980924 | 1.20 | 0.232 | -.1526487 | .6277301 |
| 3 | 1.33357 | .2313119 | 5.77 | 0.000 | .8779474 | 1.789193 |
| 4 | 2.136807 | .322162 | 6.63 | 0.000 | 1.502234 | 2.771381 |
| 5 | 5.298241 | .6567774 | 8.07 | 0.000 | 4.004564 | 6.591918 |
| TOTROOMS | .4247403 | .0771275 | 5.51 | 0.000 | .2728197 | .5766609 |
| rCOOLTYPE | | | | | | |
| 1 | 3.276034 | .1733322 | 18.90 | 0.000 | 2.934616 | 3.617453 |
| 2 | 2.006208 | .1659761 | 12.09 | 0.000 | 1.679279 | 2.333137 |
| 3 | 4.797121 | .6626675 | 7.24 | 0.000 | 3.491842 | 6.102399 |
| TOTSQFT | .0015727 | .0001185 | 13.27 | 0.000 | .0013393 | .0018061 |
| _cons | -8.641008 | .4751101 | -18.19 | 0.000 | -9.576848 | -7.705167 |

. predict double resid, residual

. svy: mean totbtucolM resid

. generate v1 = e(N)*el(e(V_srs),1,1)

. generate v2 = e(N)*el(e(V_srs),2,2)

. generate adjr2= (1-v2/v1)

. label var adjr2 "Adjusted R-Sq"

. format adjr2 %8.4f

. dlist adjr2 in 1, name(0)

1.

Adjusted R-Sq    0.5451

```
.      * unwtd regn
. regress totbtucolM CDD65 i.AIA_Zone TOTROOMS i.rCOOLTYPE TOTSQFT
```

| Source | SS | df | MS | | | |
|---|---|---|---|---|---|---|
| | | | | Number of obs | = | 12,083 |
| | | | | F(10, 12072) | = | 1462.10 |
| Model | 473520.967 | 10 | 47352.0967 | Prob > F | = | 0.0000 |
| Residual | 390967.647 | 12,072 | 32.3863194 | R-squared | = | 0.5477 |
| | | | | Adj R-squared | = | 0.5474 |
| Total | 864488.614 | 12,082 | 71.5517807 | Root MSE | = | 5.6909 |

| totbtucolM | Coef. | Std. Err. | t | P>\|t\| | [95% Conf. Interval] | |
|---|---|---|---|---|---|---|
| CDD65 | .002858 | .0001116 | 25.61 | 0.000 | .0026392 | .0030767 |
| | | | | | | |
| AIA_Zone | | | | | | |
| 2 | .4136982 | .2033062 | 2.03 | 0.042 | .0151855 | .8122109 |
| 3 | 1.665264 | .2070625 | 8.04 | 0.000 | 1.259388 | 2.071139 |
| 4 | 2.189723 | .2308973 | 9.48 | 0.000 | 1.737127 | 2.642319 |
| 5 | 5.711743 | .3738844 | 15.28 | 0.000 | 4.978869 | 6.444616 |
| | | | | | | |
| TOTROOMS | .4154039 | .0346243 | 12.00 | 0.000 | .3475347 | .4832731 |
| | | | | | | |
| rCOOLTYPE | | | | | | |
| 1 | 3.316157 | .1522256 | 21.78 | 0.000 | 3.017771 | 3.614544 |
| 2 | 2.032663 | .1715604 | 11.85 | 0.000 | 1.696377 | 2.368949 |
| 3 | 4.816733 | .480731 | 10.02 | 0.000 | 3.874424 | 5.759043 |
| | | | | | | |
| TOTSQFT | .0016957 | .0000525 | 32.29 | 0.000 | .0015927 | .0017986 |
| _cons | -9.211274 | .2457001 | -37.49 | 0.000 | -9.692886 | -8.729662 |

```
. #delimit ;
. regress TOTALBTUCOL CDD65 i.AIA_Zone TOTROOMS i.rCOOLTYPE TOTSQFT
>         w_one w_cdd65
>         w_aiazone1 w_aiazone2 w_aiazone3 w_aiazone4 w_aiazone5
>         w_totrooms
>         w_cooltype1 w_cooltype2 w_cooltype3 w_cooltype0 ;
. #delimit cr

. #delimit ;
. test w_one w_cdd65
>         w_aiazone1 w_aiazone2 w_aiazone3 w_aiazone4 w_aiazone5
>         w_totrooms w_cooltype1 w_cooltype2 w_cooltype3 w_cooltype0 ;
(output omitted)
       F( 10, 12062) =     7.25
           Prob > F =    0.0000
. #delimit cr
```

Finally, an OLS regression is run that includes the same independent variables plus the w_* variables, that is, $\mathbf{XW}$ in the extended model. This regression has adjusted $R^2 = 0.5497$, which is virtually the same as those for the weighted and unweighted regressions that do not include $\mathbf{XW}$. A joint test of whether all coefficients on $\mathbf{XW}$ are simultaneously 0 is performed with the test statement. This hypothesis is rejected, but this is due to having a large sample size—12,083 households. The fact that the adjusted $R^2$'s are essentially the same for all three of the models above seems like convincing evidence that an unweighted regression will give consistent estimates of parameters.

However, RECS is a multistage sample, and accounting for the clustering in the sample design is likely to be important when estimating SEs. The BRR replicates in RECS are constructed to account for that clustering, but neither the strata nor cluster identifiers are in the public-use file. If the standard BRR had been used where some PSUs are entirely omitted from each replicate, one approach is to create a new set of BRR weights that are 0 if a BRR weight is 0 and are 1 if a BRR weight is greater than 0. However, RECS used the Fay variation of BRR (see section 5.4.2), so no PSUs were fully dropped from any replicate. Consequently, the only choice with this dataset for an analyst who wants to obtain legitimate SEs is to use the BRR weights and to fit survey-weighted regressions.

■

The following example, which also uses replicate weights, does allow the possibility of doing an unweighted regression that accounts for strata and clusters in a sample design. As noted by Nordberg (1989), the same type of extended regression as above can be used to test whether the weights should be used in generalized linear models.

**Example 7.3: Testing use of weights in a logistic regression.** This example revisits the National Maternal and Infant Health Survey used in example 5.9. As noted in that example, the sample is a single-stage design, stratified by whether a birth was live, a fetal death, or an infant death. We fit a logistic regression to predict whether a baby had a low birthweight (less than 2500g) based on whether a mother had a previous miscarriage (`miscar`), the mother's age, whether the mother had multiple babies in the current birth (`multiple`), the sex of the child (`childsex`), and whether the mother had vaginal bleeding during the pregnancy (`vagbleed`). (The code is in `ex.7.3_svywt.test.logistic.do`.) Although the dataset does include live births, fetal deaths, and infant deaths, all records do include a birthweight. One notable feature of the dataset is that the final weight (`finwgt`) is related to having a low birthweight baby. None of the cases with `finwgt` over about 800 have low birthweight. This leads to weighted logistic regression being quite different from unweighted logistic regression.

As a first step, the **XW** variables are created. The unweighted regression has a log likelihood of $-5664.4974$ and pseudo-$R^2 = 0.0757$. In contrast, the extended model has a log likelihood of $-3183.3619$ and pseudo-$R^2 = 0.4806$. The test of whether the `w_*` predictors are simultaneously 0 is strongly rejected. These comparisons are saying that weighted regressions should be run, especially because no design information other than replicate weights is provided.

```
. use http://www.stata-press.com/data/svywt/nmihs_bs, clear
. generate one = 1
. generate w_one = finwgt*one
. generate w_miscar = finwgt*(miscar == 1)
. generate w_age = finwgt*age
. generate w_mult = finwgt*(multiple == 1)
```

```
. generate w_sex = finwgt*(childsex == 1)

. generate w_vagbleed = finwgt*(vagbleed == 1)

. generate lbw = lowbw == 1


.     * unweighted fit
. logit lbw i.miscar age i.multiple i.childsex i.vagbleed
  (output omitted)
```

| Logistic regression | | | | Number of obs | = | 9,953 |
|---|---|---|---|---|---|---|
| | | | | LR chi2(5) | = | 927.72 |
| | | | | Prob > chi2 | = | 0.0000 |
| Log likelihood = -5664.4974 | | | | Pseudo R2 | = | 0.0757 |

| lbw | Coef. | Std. Err. | z | P>|z| | [95% Conf. Interval] | |
|---|---|---|---|---|---|---|
| **miscar** | | | | | | |
| miscar | .1482997 | .0641231 | 2.31 | 0.021 | .0226208 | .2739787 |
| age | -.0128697 | .0040555 | -3.17 | 0.002 | -.0208183 | -.0049211 |
| **multiple** | | | | | | |
| multi | 2.768573 | .1243944 | 22.26 | 0.000 | 2.524764 | 3.012382 |
| 2.childsex | .1095033 | .0458828 | 2.39 | 0.017 | .0195747 | .1994319 |
| **vagbleed** | | | | | | |
| bleed | 1.075617 | .0914297 | 11.76 | 0.000 | .8964181 | 1.254816 |
| _cons | -.7944869 | .1072461 | -7.41 | 0.000 | -1.004685 | -.5842884 |

```
.     * extended model
. logit lbw i.miscar age i.multiple i.childsex i.vagbleed w_one w_miscar w_age
> w_mult w_sex w_vagbleed
  (output omitted)
```

| Logistic regression | | | | Number of obs | = | 9,953 |
|---|---|---|---|---|---|---|
| | | | | LR chi2(11) | = | 6156.82 |
| | | | | Prob > chi2 | = | 0.0000 |
| Log likelihood = -3049.946 | | | | Pseudo R2 | = | 0.5023 |

| lbw | Coef. | Std. Err. | z | P>|z| | [95% Conf. Interval] | |
|---|---|---|---|---|---|---|
| **miscar** | | | | | | |
| miscar | .0661828 | .2316478 | 0.29 | 0.775 | -.3878386 | .5202041 |
| age | -.0726981 | .0144573 | -5.03 | 0.000 | -.1010339 | -.0443623 |
| **multiple** | | | | | | |
| multi | .9578117 | .3054792 | 3.14 | 0.002 | .3590834 | 1.55654 |
| 2.childsex | -.1466341 | .1753377 | -0.84 | 0.403 | -.4902897 | .1970214 |
| **vagbleed** | | | | | | |
| bleed | .3862394 | .2841768 | 1.36 | 0.174 | -.1707369 | .9432157 |
| w_one | -.0435474 | .002712 | -16.06 | 0.000 | -.0488629 | -.038232 |
| w_miscar | .0010671 | .0015481 | 0.69 | 0.491 | -.0019672 | .0041014 |
| w_age | .0007309 | .0000993 | 7.36 | 0.000 | .0005363 | .0009255 |
| w_mult | .012213 | .0015861 | 7.70 | 0.000 | .0091043 | .0153217 |
| w_sex | -.0019207 | .0011858 | -1.62 | 0.105 | -.004245 | .0004035 |
| w_vagbleed | .0021288 | .0018194 | 1.17 | 0.242 | -.0014371 | .0056948 |
| _cons | 4.911829 | .3977518 | 12.35 | 0.000 | 4.13225 | 5.691408 |

```
Note: 2002 failures and 0 successes completely determined.
```

```
. test w_one w_miscar w_age w_mult w_sex w_vagbleed
```
(*output omitted*)
```
         chi2(  6) = 1539.60
       Prob > chi2 =    0.0000
```

The output of the extended regression includes the following: `Note: 2002 failures and 0 successes completely determined`. There are 6,902 nonlow birthweight cases in the sample. "Completely determined" means that 2,002 of those cases are predicted to be nonlow with certainty. As explained by Sribney (undated), this can be caused by an extremely strong predictor, which `finwgt` is in this example.

∎

Suppose for the sake of illustration that we decided to use an unweighted regression above. Assume that the bootstrap replicates were formed considering the strata and clusters (if any) in the design. The effects on SEs of complexities of the sample design can be accounted for by coding each nonzero bootstrap weight to 1 and retaining the zero weights as they are.

**Example 7.4: Setting weights to 1 but accounting for other design features in logistic regression.** In the `foreach` loop below, a new set of bootstrap weights equal to 1 or 0 is created from the weights in example 7.3. These `bw*` weights are then used along with full-sample weights equal to 1 (`one`) to run the same regression as above. The coefficient estimates are the same as those from the unweighted regression in example 7.3, but the SEs are somewhat smaller, presumably because the bootstrap SEs account for the efficiency of the stratified design. The code is in `ex.7.4_svywt.wts.one.do`.

```
. foreach i of numlist 1/1000 {
  2.        generate bw`i´ = 0
  3.        replace bw`i´ = 1 if bsrw`i´ > 0
  4. }
. svyset [pweight=one], bsrweight(bw1 - bw1000) vce(bootstrap) mse

. svy: logit lbw i.miscar age i.multiple i.childsex i.vagbleed
  (output omitted )
```

Survey: Logistic regression                     Number of obs    =       9,953
                                                Population size   =       9,953
                                                Replications      =       1,000
                                                Wald chi2(5)      =     1050.96
                                                Prob > chi2       =      0.0000

| lbw | Observed Coef. | Bstrap * Std. Err. | z | P>\|z\| | [95% Conf. Interval] | |
|---|---|---|---|---|---|---|
| miscar | | | | | | |
| miscar | .1482997 | .0508939 | 2.91 | 0.004 | .0485494 | .24805 |
| age | -.0128697 | .0031937 | -4.03 | 0.000 | -.0191293 | -.0066101 |
| | | | | | | |
| multiple | | | | | | |
| multi | 2.768573 | .0970245 | 28.53 | 0.000 | 2.578409 | 2.958738 |
| 2.childsex | .1095033 | .0356054 | 3.08 | 0.002 | .039718 | .1792886 |
| | | | | | | |
| vagbleed | | | | | | |
| bleed | 1.075617 | .0705543 | 15.25 | 0.000 | .9373331 | 1.213901 |
| _cons | -.7944869 | .0830309 | -9.57 | 0.000 | -.9572245 | -.6317493 |

# 8    Quality of survey weights

Survey quality is an ongoing topic of conversation. How is quality best defined? How might we measure quality in a comparable and transparent way? Biemer and Lyberg (2003), for example, give several definitions of quality used by organizations around the world. This topic is discussed from a total error perspective in texts such as Biemer et al. (2017) where the authors review study protocols and adjustments to improve quality. Few sources, however, delve into quality of survey weights specifically.

Kolczynska, Zielinski, and Powalko (2016) herald the importance of high-quality weights for combining data from multiple surveys; their definition of quality, however, appears to be specific to their situation only. Kalton and Flores-Cervantes (2003) recommend that statisticians verify the assumptions underlying models used in the weight adjustment process. They also recommend careful evaluation of the auxiliary information used in these models, along with the structure of the models themselves. For example, marginal-only models such as those used with raking (iterative proportional fitting) may be preferred, but researchers should evaluate the inclusion of interaction affects (say, through methods, such as a tree-based algorithm, discussed in section 3.3). Valliant et al. (2013, chap. 18) discuss a few quality checks for the weights, such as evaluating frame information and base weights just after sampling occurs and comparing weight sums against population controls. Consequently, more guidance is needed.

The quality of survey weights is only as good as the procedures put in place (assurance) and steps to ensure those procedures were implemented correctly (control). In the sections below, we discuss several aspects of quality assurance (QA) and quality control (QC) for weighting, building on work by Valliant et al. (2013). We begin with recommendations for QA and QC in the planning phase of the study (section 8.1), followed by checks on the base weights (section 8.2). Next, we provide recommendations for ensuring quality of the inputs to the weighting process such as sample status codes (section 8.3) and calibration control totals (section 8.5). We provide suggested QC checks for the adjusted weights in section 8.6. Not only do we discuss evaluating the weights themselves, but we also recommend that you examine the weights in action by producing population estimates for a set of characteristics important to your survey (section 8.7). We conclude this chapter in section 8.8 with an emphasis on documentation that should accompany the weights upon completion of the quality review.

# 8.1   Design and planning stage

Extensive focus and resources are used in the development of a study design. Components of the study design include the sampling plan, data collection protocols, analysis plans, and the like. The same attention should be applied to the quality of the survey weights even before the sample is drawn. Below, we provide key aspects of this critical period for the project that can complicate the weighting process.

- *Definition of the target population.* This seems simple, yet rarely is when you are staring down at a sampling frame or a set of disposition codes. Members of the target populations must be clearly defined so that if possible nonmembers (that is, ineligibles) can be removed from the sampling frame prior to selection. The explicit definition is even more important post-data collection to identify those sample cases excluded from certain weight adjustments. Mapping disposition codes to the four AAPOR (2016) categories discussed in section 1.6 when there is time to reflect upon the decisions can ease the process as data start to roll in and results are required soon after.

- *Weighting specifications.* Upon completion of the sampling design, researchers should develop written specifications for how the sample is drawn and the base weights constructed. This information is useful not only for assuring quality but is also necessary for documentation. These specifications should be thorough and include measures to determine if the specifications were followed. We discuss specific QC steps in the subsequent sections.

- *Weighting programs.* If not already standard, the project team should create protocols for how the sampling and weighting programs are developed, maintained, documented (internal and external to the program), and stored if persons initially completing the tasks are replaced with new project team members. In example 8.1, we provide an example of initial documentation included in a Stata program referred to as a header. A general discussion of these issues is provided in Valliant et al. (2013, sec. 18.7) with further details not covered here.

- *Official review and sign-off.* A second pair of eyes is always useful. With this in mind, the project team should designate someone with the appropriate skills to provide an official review. The reviewer is a person other than those deeply involved in the work. This requires a link between the well-documented sampling and weight specifications, the statistical programs, and the results from the QC steps to facilitate the review. The level of review may include the re-creation of a complicated statistical program or only a review of the results (with potentially extra QC evaluations requested for good measure). Official sign-off is obtained only after the reviewer and reviewee come to agreement that the weighting process was a success. Including the cost of these "extra eyes" is, naturally, important at the budgeting stage.

**Example 8.1: Example Stata program header.**

```
****************************************************************************
* PROGRAM: 2_Check_BaseWts.do                                             *
* DATE:    5 March 2017                                                   *
* AUTHOR:  L. Kish                                                        *
* PURPOSE: Verify base weights calculated in 1_Sample.do                  *
* SPECS:   \\ ...\Weight_specs\Check_BaseWts.docx                         *
* REVISED: 10 March 2017 (L. Kish)                                        *
*              Added unequal weighting effect calculation.                *
****************************************************************************
```

■

## 8.2  Base weights

Protocols have been established. The sampling frame(s) have been evaluated. The sample has been selected after verifying the sampling specifications and addressing any selection probabilities that are greater than one. All of this happens before data collection has even started. Now your focus moves to the survey weights.

Immediately following sample selection, the base weights should be calculated. We recommend a few quality checks for this first step. For example, verify that the sum of the selection probabilities across the entire frame equals the desired sample size. In a pps sample, for example, with mos equal to $x_i$ for unit $i$, $\sum_U nx_i/(N\overline{x}_U) = n \sum_U x_i/(N\overline{x}_U) = n$. After calculating the base weights by inverting the probabilities (section 2.1), verify that the sum of the weights is (approximately) equal to the population total number of units on your sampling frame. Rounding error associated with weight calculation will affect the closeness. But, with some methods like pps sampling (section 1.4), the sum of the sample weights will be somewhat larger than $N$. This comparison should be implemented for any important subdomain described in the analysis plan. Because they are relevant after each weight adjustment, additional weight statistics of importance are discussed in section 8.6.

**Master database**. A helpful accounting practice we have found useful in our work is a master (sample) database. This Stata file contains all relevant case-level information needed to produce and verify the final weights and enable the calculation of appropriate population estimates. The master database is initially populated with the sampled records and one or more unique case-level identifiers (IDs) such as internal IDs and those made available to the public. The following are example variables merged on to the database using the unique ID(s) before the sample is released for data collection:

- Characteristics associated with stratification

- One or more unique stratum IDs

- One or more unique cluster IDs

- Sample size per stratum

- Components of the record-specific mos (for example, composite size measure)
- Sum of the mos by stratum
- Probability of selection
- Base weight
- Sample frame indicator (if multiple frames are used)
- Additional variables important for sample monitoring or analysis

For example, two or more unique stratum and cluster IDs are needed if your design has at least three stages of selection with samples drawn from strata defined for more than one stage.

As data collection progresses, information is added as needed to account for the sample status. For example, indicators of "sampled but not released" and "eligible but not sampled for a nonresponse follow-up" (that is, phase 2) may be relevant for your study and paramount to the correct calculation of the weights. Final disposition codes and associated AAPOR categories are appended as the status for the sampled cases is finalized as discussed in the next section.

## 8.3   Data editing and file preparation

Many surveys today are conducted in an electronic medium that includes verification checks on responses during the interview. Study participants are prompted to verify or correct any inconsistent or out-of-range responses. The intended result is a clean dataset without "catchable" errors (that is, excluding undetectable measurement errors). However, it is wise to review the data, especially items related to classifying the final disposition for the case and to constructing the weights. We discuss both item sets below.

### Disposition codes

The weighting specifications should include a one-to-one mapping of the final data collection disposition (status) codes to the four AAPOR categories. As discussed in section 1.6, the AAPOR categories are eligible respondent (ER), eligible nonrespondent (ENR), unknown eligibility (UNK), and ineligible (IN). This task must be completed after the disposition codes have been verified and before adjustments such as unknown eligibility (section 2.2) are calculated.

The mapping should cover all disposition codes used for the survey and, once constructed, should be verified with actual data. Codes for sample units known (or assumed) to be members of the target population are mapped to either ER or ENR depending on their response status. Codes linked to units verified to fall outside the target population are mapped to IN. All others are classified as UNK.

The distinction between the eligible categories is not always clear cut. For example, the classification of a partially completed interview into either ER or ENR is ideally resolved in the planning stage based on whether a unit completed critical items needed for analyses. However, researchers may make this determination after examining all or a portion of the interviews.

Consequently, the examination of the disposition code along with interview responses is an important first step toward the creation of high-quality survey weights. Then the mapping of disposition code to AAPOR response category can be verified.

## Variables relevant for weighting

Examination of the data does not stop once the response categories are set. Next, researchers should examine all items that may be included in the weighting process using the weighting specifications as a guide. For example, if weights are to be calibrated to population totals for gender and race, you will want to impute values for these items that were not obtained during the interview. This also includes any discrepant responses that were not corrected during the interview, because you may choose to replace this information with an imputed value.

# 8.4 Models for nonresponse and calibration

Throughout this text, we have demonstrated the power of models to produce survey weights. However, if left unchecked, these models may do more harm than good with respect to the mean square error (MSE = variance + bias squared) of your estimates. Below we comment on a few aspects of QA and QC for models.

Kalton and Flores-Cervantes (2003), as previously documented, recommend that researchers evaluate possible interaction terms instead of using a less-complex main effects only response model. Random-forest (section 3.3.2) and boosting (section 3.3.3) techniques use iterative procedures to identify such interactions and possible main effects for the model. Both techniques have been shown to have better predictive properties when compared with the noniterative CART method (section 3.3.1).

Predictive properties of the response model can be evaluated with a hold-out (or test) sample for a model developed with a training dataset (see section 3.2). Kalton and Flores-Cervantes (2003) also suggest an evaluation of model assumptions. Many sources exist for conducting model diagnostics including Heeringa, West, and Berglund (2017), Li and Valliant (2011), Liao and Valliant (2012), and Li and Valliant (2015).

Lastly, models are not guaranteed to converge to a solution for either nonresponse or calibration. Therefore, we recommend scrutinizing the program logs (a critical step for any type of programming!) to determine convergence and the SEs for the model coefficients. Relatively large SEs are indicative of model instability and the need for further examination.

## 8.5   Calibration totals

Chapter 4 stressed the importance of weight calibration as a final step in the process. Source or sources of control totals are identified that correspond with the information collected in your survey and are relevant to your target population. However, when the data collection instrument is designed, following this advice is critical:

> **Warning.**   If external sources, like a large government-sponsored survey, are used for control totals, the technique used to collect the calibration variables from sample units must match the one used in the external source. For example, if other race is used as calibration variables, your survey should ask the other race questions with exactly the same wording as the external source. Departing from that wording will cause your sample estimates of other race totals to be incomparable with the external controls.

A few QC checks and associated decisions are warranted before you construct the calibration weights.

First, we recommend that you calculate the sum of the input weights corresponding to your control totals. If the weight sums (an estimate of the population size) differ greatly from the control total values, analysts should examine the construction of the input weights to identify problems or evaluate the alignment of the survey and the control total sources. If the control totals are estimated internally (for example, removal of estimated ineligibles from population projections), then you should carefully review their construction. If your survey poorly covers the target population, calibration may well lead to substantial increases in weights in order to hit the control totals. In that case having input-weighted estimates that are considerably less than control totals is expected. But, evaluating whether large weight adjustments are consistent with the anticipated coverage of the survey is important.

Second, the calibration model should be populated with covariates that are associated with the key items in your survey. As discussed in Kalton and Flores-Cervantes (2003), analysts should evaluate the weight adjustment models for important interaction terms if the corresponding population totals are available. Regarding the use of terms in the calibration model that are not significantly associated with the key survey items, there are two schools of thought: 1) If the covariate is substantively meaningful for your analyses and may be significantly associated with other survey items, then the covariate should be included. The additional variability in the survey weights that may occur is less relevant than mean square error reduction for those other survey items. 2) All insignificant covariates should be excluded from the model to limit weight variability. In practice, however, the number of available covariates is typically limited, so the choice of either approach will not demonstrate a meaningful difference. However, comparing the weights from both models using criteria discussed in section 8.6 may provide insight on the best path to follow.

## 8.6 Weighting checks

The following checklist serves as a guide for some statistics that you may examine before and after each adjustment is calculated for analytically meaningful domains. Large, unexpected changes may indicate a problem with the calculation or with the approach (for example, small weighting-class cells resulting in large adjustments). The definition of "large" is somewhat subjective if not already defined by your organization. This list includes a core set of statistics but is likely not comprehensive.

- *Record count.* The number of records included at each step of the weighting process (see figure 1.2) should be known and verified. Developers should also verify the number of records with a zero or a positive weight for both input and output weights. The number of records entering a step should equal the number of records in files exiting the step.

- *Weight sum.* The weight sum serves as an estimate of a population size. The weight sums by domain should be compared before and after an adjustment is applied and compared against known population values. A very important example is to compare the sum of the final weights against the control total to verify the calibration adjustment.

- *Range.* Large shifts in the minimum and maximum weights and weight adjustments may indicate the need to examine bounds set for the adjustments. This is also used to identify negative weights or potential outliers. We discuss the identification of outliers below.

- *Percentiles.* Unexpected shifts in the distribution of the weights and weight adjustments are another indication of trouble or the need to modify the methodology. Several sources, including Westat (2007), suggest examining the set of percentiles 1, 5, 10, 25, 50, 75, 90, 95 and 99.

- *Variability.* Excessive weight variability can lower precision of your estimates. One measure discussed in section 4.5 is the unequal weighting effect (UWE), which determines the variability of your weighting system relative to a stratified simple random sample with a proportional allocation. Deliberate over- or undersampling for analytic needs will naturally inflate the UWE and is of no worry. UWEs are very effective in the before- and after-adjustment comparison. For example, a three-fold increase in the UWE after implementing a weighting step might suggest the need for corrective action.

- *Outliers.* Excessively large or small weights relative to the rest of the respondent set may need to be addressed if the precision of your estimates is unduly impacted. Typically, weights that fall outside of a specified range are evaluated for outlier status. One commonly used cutoff value is the median weight $\pm\, 3 \times$ interquartile range (IQR), where IQR = 75th percentile − 25th percentile (Potter 1988). Another cutoff value could be the mean weight $\pm\, 3 \times$ the standard deviation of the weights. Once the weights are verified as correct, the analyst may 1) tighten bounds on the weight adjustments, 2) trim the weights (a topic not covered in this text), or 3) leave the weights enough alone.

## 8.7   Analytic checks

In addition to statistics on the weights, evaluating population estimates for a few of the key items from your survey (identified as such in your analysis plan) is another good approach. We suggest the following analyses:

- *Coefficient of variation.* The cv, also known as an estimated relative SE, is calculated as the SE divided by the point estimate. The square of the cv is the relvar (relative variance) discussed previously. Some organizations dictate that percent cv ($100 \times$ cv) greater than a specified level either should be suppressed from publication or noted as unstable. For example, the U.S. Centers for Disease Control generally sets this level at 30 percent. If the cv for the key survey items exceeds this level, then additional efforts may be needed to reduce the SEs if possible.

- *Design effects.* Like the cv, design effects (deffs) are another good tool if there are such estimates from, say, a prior round of the survey with which to compare. Large changes in deff or cv for the same design may indicate a need to tighten bounds on, for example, one or more nonresponse adjustments.

- *Standard errors.* For surveys with replicate weights, estimates (point and SEs) calculated with the full-sample weight (linearization) should be evaluated against the replicate estimates. Large differences may indicate a need to evaluate nonresponse or calibration adjustments appropriate for the full sample that proved unstable for one or more replicates.

- *Nonresponse bias.* Nonresponse bias was discussed in chapter 3. Large differences in estimates calculated separately for ER and ENR indicate that the nonresponse weight adjustments were not completely effective in reducing bias to negligible levels at least for that variable. However, because we require information for both the ERs and ENRs, this evaluation is typically limited to variables from the sampling frame. Unless this information is associated with the key items from your survey, this evaluation is only suggestive of nonresponse bias throughout.

- *Comparison with other surveys.* Differences in estimates from your survey and those from other surveys may indicate a need to include more calibration control totals to minimize any differences. However, this is a lively area of research with limited guidance on methods to improve estimate comparability.

## 8.8 Analysis file and documentation

We conclude this brief chapter on the importance of QA and QC for weights with comments on the final analysis file and associated documentation.

### Analysis file

Before the analysis file(s) is released to others for further analyses, a few QC checks are required. 1) Compare the total number of records with a positive weight against the number of respondents designated in your master database. In general, only respondent records are included in the analysis file. This QC check should be conducted for the respondent-sample weight and the replicate weights. 2) Verify that the analysis file (or the master database if required for delivery) contains all components needed to regenerate the final analysis weight. This affords a quality review by others if questions on the components arise. 3) Though not explicitly associated with weight calculation, verify that the IDs included in the analysis file(s) merge uniquely with the master database if you need to redeliver the file(s) for any reason.

### Documentation

Many surveys require documentation to support release of an analytic data file(s). This external documentation is intended for distribution to the survey sponsor and possibly to the public. It differs from the internal documentation used by the organization conducting the survey. The internal documentation includes specification memos and programs and must be precise enough to allow details to be redone if there is a need to retrace your steps because of errors or changes in the survey design.

The external documentation will include details of the sample design and information on the base weights and associated adjustments. This documentation builds on the sampling plan and weighting specifications developed during the planning phase along with the QA and QC protocols implemented and summary statistics generated during the final QC process. This documentation is the critical communication vehicle to ensure appropriate data use. It also provides needed information to advertise the weights (and the study in its entirety) as a high-quality initiative and to allow comparisons with other surveys.[1]

---

1. See http://www.aapor.org/Transparency_Initiative.htm.

# References

AAPOR. 2016. Standard Definitions: Final Dispositions of Case Codes and Outcome Rates for Surveys. Technical report, American Association for Public Opinion Research, Oakbrook Terrace, IL. http://www.aapor.org/AAPOR_Main/media/publications/Standard-Definitions20169theditionfinal.pdf.

Alvarez, R. M., R. P. Sherman, and C. Van Beselaere. 2003. Subject acquisition for web-based surveys. *Political Analysis* 11: 23–43.

Baker, R., S. J. Blumberg, J. M. Brick, M. P. Couper, M. Courtright, J. M. Dennis, D. Dillman, M. R. Frankel, P. Garland, R. M. Groves, C. Kennedy, J. Krosnick, P. J. Lavrakas, S. Lee, M. Link, L. Piekarski, K. Rao, R. K. Thomas, and D. Zahs. 2010. AAPOR report on online panels. *Public Opinion Quarterly* 74: 711–781.

Baker, R., J. M. Brick, N. A. Bates, M. Battaglia, M. P. Couper, J. A. Dever, K. J. Gile, and R. Tourangeau. 2013a. Report of the AAPOR Task Force on Non-Probability Sampling. Technical report, American Association for Public Opinion Research, Oakbrook Terrace, IL. http://www.aapor.org/AAPOR_Main/media/MainSiteFiles/NPS_TF_Report_Final_7_revised_FNL_6_22_13.pdf.

———. 2013b. Summary report of the AAPOR task force on non-probability sampling. *Journal of Survey Statistics and Methodology* 1: 90–143.

Bauer, E., and R. Kohavi. 1999. An empirical comparison of voting classification algorithms: Bagging, boosting, and variants. *Machine Learning* 36: 105–139.

Bethlehem, J. 2010. Selection bias in web surveys. *International Statistical Review* 78: 161–188.

Biemer, P. P., E. de Leeuw, S. Eckman, B. Edwards, F. Kreuter, L. E. Lyberg, N. C. Tucker, and B. T. West, ed. 2017. *Total Survey Error in Practice*. Hoboken, NJ: Wiley.

Biemer, P. P., and L. E. Lyberg. 2003. *Introduction to Survey Quality*. Hoboken, NJ: Wiley.

Binder, D. A. 1983. On the variances of asymptotically normal estimators from complex surveys. *International Statistical Review* 51: 279–292.

Binder, D. A., and G. Roberts. 2009. Design- and model-based inference for model parameters. In Vol. 29B of *Handbook of Statistics: Sample Surveys: Inference and Analysis*, ed. D. Pfeffermann and C. R. Rao, chap. 24, 33–54. Amsterdam: Elsevier.

Breidt, F. J., and J. D. Opsomer. 2007. Comment: Struggles with survey weighting and regression modeling. *Statistical Science* 22: 168–170.

Breiman, L. 2001. Random forests. *Machine Learning* 45: 5–32.

Breiman, L., J. H. Friedman, R. Olshen, and C. Stone. 1984. *Classification and Regression Trees*. Belmont, CA: Wadsworth.

Brick, J. M., I. Flores-Cervantes, S. Lee, and G. Norman. 2011. Nonsampling errors in dual frame telephone surveys. *Survey Methodology* 37: 1–12.

Brown, D., J. A. Dever, L. B. Squiers, and E. M. Augustson. 2016. Recruiting a survey sample online: Detecting and preventing fraud. AAPOR 71st Annual Conference, Austin, TX.

Chipman, H. A., E. I. George, and R. E. McCulloch. 2010. BART: Bayesian Additive Regression Trees. *Annals of Applied Statistics* 4: 266–298.

Cochran, W. G. 1977. *Sampling Techniques*. 3rd ed. New York: Wiley.

Couper, M. P. 2000. Web surveys: A review of issues and approaches. *Public Opinion Quarterly* 64: 464–494.

———. 2013. Is the sky falling? New technology, changing media, and the future of surveys. *Survey Research Methods* 7: 145–156.

D'Agostino, R. B., Jr. 1998. Propensity score methods for bias reduction for the comparison of a treatment to a non-randomized control group. *Statistics in Medicine* 17: 2265–2281.

Defense Manpower Data Center. 2004. May 2004 Status Of Forces Survey Of Reserve Component Members: Administration, Datasets, and Codebook. Technical Report No. 2004-013, Defense Manpower Data Center, Arlington, VA.

Dever, J. A. 2008. Sampling Weight Calibration with Estimated Control Totals. PhD thesis, University of Maryland. https://drum.lib.umd.edu/handle/1903/8815.

Dever, J. A., A. Rafferty, and R. Valliant. 2008. Internet surveys: Can statistical adjustments eliminate coverage bias? *Survey Research Methods* 2: 47–62.

Dever, J. A., and R. Valliant. 2010. A comparison of variance estimators for poststratification to estimated control totals. *Survey Methodology* 36: 45–56.

———. 2014. Estimation with non-probability surveys and the question of external validity. In *Proceedings of Statistics Canada Symposium 2014*. Statistics Canada.

———. 2016. General regression estimation adjusted for undercoverage and estimated control totals. *Journal of Survey Statistics and Methodology* 4: 289–318.

Deville, J.-C., and Y. Tillé. 2004. Efficient balanced sampling: The cube method. *Biometrika* 91: 893–912.

Ding, H., C. L. Black, S. Ball, S. Donahue, R. V. Fink, W. W. Williams, E. D. Kennedy, C. B. Bridges, P.-J. Lu, K. E. Kahn, A. K. Dean, L. A. Grohskopf, I. B. Ahluwalia, R. Devlin, C. DiSogra, D. K. Walker, and S. M. Greby. 2015. Influenza vaccination coverage among pregnant women—United States, 2014–15 influenza season. *Morbidity and Mortality Weekly Report* 64: 1000–1005.

Dong, Q., M. R. Elliott, and T. E. Raghunathan. 2014. A nonparametric method to generate synthetic populations to adjust for complex sampling design features. *Survey Methodology* 40: 29–46.

DuMouchel, W. H., and G. J. Duncan. 1983. Using sample survey weights in multiple regression analyses of stratified samples. *Journal of the American Statistical Association* 78: 535–543.

Eckman, S., K. Himelein, and J. A. Dever. Forthcoming. Sample Designs Using GIS Technology for Household Surveys in the Developing World. In *New Ideas in Sampling for Surveys in the Developing World*, ed. T. Johnson, B.-E. Pennell, I. Stoop, and B. Dorer, chap. 5. New York: Wiley.

Efron, B. 1982. *The Jackknife, the Bootstrap and Other Resampling Plans*. Philadelphia: SIAM [Society for Industrial and Applied Mathematics].

Elliott, M. R. 2009. Combining data from probability and non-probability samples using pseudo-weights. *Survey Practice*. https://surveypractice.files.wordpress.com/2009/08/elliott.pdf.

Elliott, M. R., and R. Valliant. 2017. Inference for nonprobability samples. *Statistical Science* 32: 249–264.

Ericson, W. A. 1969. Subjective Bayesian models in sampling finite populations. *Journal of the Royal Statistical Society, Series B* 31: 195–233.

Fay, R. E. 1984. Some properties of estimates of variance based on replication methods. In *Proceedings of the Survey Research Methods Section*, 495–500. Washington, DC: American Statistical Association.

File, T., and C. Ryan. 2014. Computer and Internet Use in the United States: 2013. Technical Report ACS-28, U.S. Census Bureau. http://www.census.gov/content/dam/Census/library/publications/2014/acs/acs-28.pdf.

Flores-Cervantes, I., and J. M. Brick. 2014. California Health Interview Survey, CHIS 2011–2012 Methodology Series, Report 1: Sample Design. Technical report, Los Angeles, CA: UCLA Center for Health Policy Research.

Flores-Cervantes, I., G. Norman, and J. M. Brick. 2014. California Health Interview Survey, CHIS 2011–2012 Methodology Series, Report 5: Weighting And Variance Estimation. Technical report, Los Angeles, CA: UCLA Center for Health Policy Research.

Folsom, R. E., F. J. Potter, and S. R. Williams. 1987. Notes on a Composite Size Measure for Self-weighting Samples in Multiple Domains. In *Proceedings of the Survey Research Methods Section*. Washington, DC: American Statistical Association.

Friedman, J. H. 2001. Greedy function approximation: A gradient boosting machine. *Annals of Statistics* 29: 1189–1232.

Friedman, J. H., T. Hastie, and R. J. Tibshirani. 2000. Additive logistic regression: A statistical view of boosting. *Annals of Statistics* 28: 337–407.

Fuller, W. A. 1984. Least squares and related analyses for complex survey designs. *Survey Methodology* 10: 97–118.

————. 2009. *Sampling Statistics*. Hoboken, NJ: Wiley.

Gelman, A. 2007. Struggles with survey weighting and regression modeling. *Statistical Science* 22: 153–164.

GfK. 2013. Knowledgepanel Design Summary. http://www.knowledgenetworks.com/knpanel/docs/knowledgepanel(R)-design-summary-description.pdf [Accessed 27 Sep 2017].

Ghosh, M. 2009. Bayesian Developments in Survey Sampling. In Vol. 29B of *Handbook of Statistics: Sample Surveys: Inference and Analysis*, ed. D. Pfeffermann and C. R. Rao, chap. 29, 153–187. Amsterdam: Elsevier.

Ghosh, M., and G. Meeden. 1997. *Bayesian Methods for Finite Population Sampling*. London: Chapman & Hall.

Gile, K. J., and M. S. Handcock. 2010. Respondent-driven sampling: An assessment of current methodology. *Sociological Methodology* 40: 285–327.

————. 2015. Network model-assisted inference from respondent-driven sampling data. *Journal of the Royal Statistical Society, Series A* 178: 619–639.

Gile, K. J., L. G. Johnston, and M. J. Salganik. 2015. Diagnostics for respondent-driven sampling. *Journal of the Royal Statistical Society, Series A* 178: 241–269.

Gonzalez, J. F., Jr., N. Krauss, and C. Scott. 1992. Estimation in the 1988 National Maternal and Infant Health Survey. *Proceedings of the Section on Statistics Education* 343–348.

Gosnell, H. F. 1937. How accurate were the polls? *Public Opinion Quarterly* 1: 97–105.

Gould, W., and J. Pitblado. 2017. How large should bootstrapped samples be relative to the total number of cases in the dataset? http://www.stata.com/support/faqs/statistics/bootstrapped-samples-guidelines/ [Accessed 27 Sep 2017].

Hahs-Vaughn, D. L. 2005. A primer for using and understanding weights with national datasets. *Journal of Experimental Education* 73: 221–248.

Hansen, M. H., W. N. Hurwitz, and W. G. Madow. 1953. *Sample Survey Methods and Theory, Volume I: Methods and Applications.* New York: Wiley.

Harris Interactive. 2014. The Harris Global Omnibus: Frequently Asked Questions. http://client.harrisinteractive.co.uk/Vault/Files/ HI_UK_Global_Omnibus_Research.pdf [Accessed 27 Sep 2017].

Haziza, D., and E. Lesage. 2016. A discussion of weighting procedures for unit nonresponse. *Journal of Official Statistics* 32: 129–145.

Heckathorn, D. D. 1997. Respondent-driven sampling: A new approach to the study of hidden populations. *Social Problems* 44: 174–199.

Heeringa, S. G., B. T. West, and P. A. Berglund. 2017. *Applied Survey Data Analysis.* 2nd ed. Boca Raton, FL: CRC Press.

Henry, K. A., and R. Valliant. 2015. A design effect measure for calibration weighting in single-stage samples. *Survey Methodology* 41: 315–331.

Hoeting, J. A., D. Madigan, A. E. Raftery, and C. T. Volinsky. 1999. Bayesian model averaging: A tutorial. *Statistical Science* 14: 382–401.

Hothorn, T., K. Hornik, C. Strobl, and A. Zeileis. 2016. *party: A laboratory for recursive partitioning.* R package version 1.0-25. http://CRAN.R-project.org/package=party.

Iannacchione, V. G. 2011. The changing role of address-based sampling in survey research. *Public Opinion Quarterly* 75: 556–575.

James, G., D. Witten, T. Hastie, and R. J. Tibshirani. 2013. *An Introduction to Statistical Learning: With Applications in R.* New York: Springer.

Jenkins, S. P. 2005. samplepps: Stata module to draw a random sample with probabilities proportional to size. Statistical Software Components S454101, Department of Economics, Boston College. https://ideas.repec.org/c/boc/bocode/s454101.html.

Judkins, D. R. 1990. Fay's method of variance estimation. *Journal of Official Statistics* 6: 223–239.

Kalton, G., and I. Flores-Cervantes. 2003. Weighting Methods. *Journal of Official Statistics* 19: 81–97.

Kalton, G., and D. S. Maligalig. 1991. A comparison of methods of weighting adjustment for nonresponse. *Proceedings of the U.S. Bureau of the Census, Annual Research Conference* 409–428.

Kennedy, C., A. Mercer, S. Keeter, N. Hatley, K. McGeeney, and A. Gimenez. 2016. Evaluating online nonprobability surveys: Vendor choice matters; widespread errors found for estimates based on blacks and Hispanics. http://www.pewresearch.org/2016/05/02/evaluating-online-nonprobability-surveys/ [Accessed 27 Sep 2017].

Kim, J. J., J. Li, and R. Valliant. 2007. Cell collapsing in poststratification. *Survey Methodology* 33: 139–150.

Kim, J. K., and J. Shao. 2014. *Statistical Methods for Handling Incomplete Data*. Boca Raton, FL: CRC Press.

Kish, L. 1965. *Survey Sampling*. New York: Wiley.

Kish, L., and M. R. Frankel. 1974. Inference from complex samples. *Journal of the Royal Statistical Society, Series B* 36: 1–37.

Kohler, U., and F. Kreuter. 2012. *Data Analysis Using Stata*. 3rd ed. College Station, TX: Stata Press.

Kohut, A., S. Keeter, C. Doherty, M. Dimock, and L. Christian. 2012. Assessing the representativeness of public opinion surveys. Technical report, Pew Research Center, For The People & The Press, Washington, DC. http://www.people-press.org/2012/05/15/assessing-the-representativeness-of-public-opinion-surveys/.

Kolczynska, M., M. W. Zielinski, and P. Powalko. 2016. Survey Weights as Indicators of Data Quality. Data Harmonization Newsletter. https://dataharmonization.org/2016/08/25/survey-weights-as-indicators-of-data-quality/.

Kolenikov, S. 2010. Resampling variance estimation for complex survey data. *Stata Journal* 10: 165–199.

———. 2014. Calibrating survey data using iterative proportional fitting (raking). *Stata Journal* 14: 22–59.

Korn, E. L., and B. I. Graubard. 1999. *Analysis of Health Surveys*. New York: Wiley.

Kott, P. S. 1999. Some problems and solutions with a delete-a-group jackknife. In *Federal Committee on Statistical Methodology Research Conference, Vol. 4*, 129–135. Washington, DC: U.S. Bureau of the Census.

———. 2001. The delete-a-group jackknife. *Journal of Official Statistics* 17: 521–526.

Kovar, J. G., and P. J. Whitridge. 1995. Imputation of business survey data. In *Business Survey Methods*, ed. B. G. Cox, D. A. Binder, B. N. Chinnappa, A. Christianson, M. J. Colledge, and P. S. Kott, chap. 22. New York: Wiley.

Kreuter, F., M. P. Couper, and L. E. Lyberg. 2010. The use of paradata to monitor and manage survey data collection. In *Proceedings of the Survey Research Methods Section*, 282–296. American Statistical Association.

Kreuter, F., and K. Olson. 2011. Multiple auxiliary variables in nonresponse adjustment. *Sociological Methods and Research* 40: 311–332.

Kreuter, F., S. Presser, and R. Tourangeau. 2008. Social desirability bias in CATI, IVR, and web surveys: The effects of mode and question sensitivity. *Public Opinion Quarterly* 72: 847–865.

Krewski, D., and J. N. K. Rao. 1981. Inference from stratified samples: Properties of the linearization, jackknife, and balanced repeated replication methods. *Annals of Statistics* 9: 1010–1019.

Lee, S., and R. Valliant. 2009. Estimation for volunteer panel web surveys using propensity score adjustment and calibration adjustment. *Sociological Methods and Research* 37: 319–343.

Lester, W. 1998. 'Dewey Defeats Truman' Disaster Haunts Pollsters. http://articles.latimes.com/1998/nov/01/news/mn-38174. [Accessed 27 Sep 2017].

Levy, P. S., and S. Lemeshow. 2008. *Sampling of Populations: Methods and Applications*. 4th ed. New York: Wiley.

Li, J., and R. Valliant. 2011. Linear regression influence diagnostics for unclustered survey data. *Journal of Official Statistics* 27: 99–119.

———. 2015. Linear regression diagnostics in cluster samples. *Journal of Official Statistics* 31: 61–75.

Liao, D., and R. Valliant. 2012. Condition indexes and variance decompositions for diagnosing collinearity in linear model analysis of survey data. *Survey Methodology* 38: 189–202.

Little, R. J. A. 1982. Models for nonresponse in sample surveys. *Journal of the American Statistical Association* 77: 237–250.

———. 2004. To model or not to model? Competing modes of inference for finite population sampling. *Journal of the American Statistical Association* 99: 546–556.

Little, R. J. A., and D. B. Rubin. 2002. *Statistical Analysis with Missing Data*. 2nd ed. Hoboken, NJ: Wiley.

Little, R. J. A., and S. Vartivarian. 2003. On weighting the rates in non-response weights. *Statistics in Medicine* 22: 1589–1599.

Lohr, S. L. 2010. *Sampling: Design and Analysis*. 2nd ed. Boston, MA: Brooks/Cole.

Lohr, S. L., and J. M. Brick. 2014. Allocation for dual frame telephone surveys with nonresponse. *Journal of Survey Statistics and Methodology* 2: 388–409.

Lohr, S. L., and J. N. K. Rao. 2006. Estimation in multiple-frame surveys. *Journal of the American Statistical Association* 101: 1019–1030.

Long, J. S., and L. H. Ervin. 2000. Using heteroscedasticity consistent standard errors in the linear regression model. *American Statistician* 54: 217–224.

Lu, W. W., J. M. Brick, and R. R. Sitter. 2006. Algorithms for constructing combined strata variance estimators. *Journal of the American Statistical Association* 101: 1680–1692.

Lynn, P., ed. 2009. *Methodology of Longitudinal Surveys*. Chichester, UK: Wiley.

McCarthy, P. J. 1969. Pseudo-replication: Half samples. *Review of the International Statistical Institute* 37: 239–264.

McCarthy, P. J., and C. B. Snowden. 1985. The bootstrap and finite population sampling. *Vital and Health Statistics, Series 2* 95: 85–1369.

Mendelson, J. 2014a. ppschromy: Stata module to draw sample with probability proportionate to size, using Chromy's method of sequential random sampling. Statistical Software Components S457896, Department of Economics, Boston College. https://ideas.repec.org/c/boc/bocode/s457896.html.

———. 2014b. ppssampford: Stata module to draw sample with probability proportionate to size, without replacement, using Sampford's method. Statistical Software Components S457945, Department of Economics, Boston College. https://ideas.repec.org/c/boc/bocode/s457945.html.

Nadimpalli, V., D. R. Judkins, and A. Chu. 2004. Survey calibration to CPS household statistics. *Proceedings of the Survey Research Methods Section* 4090–4094.

Newson, R. 2014. rsource: Stata module to run R from inside Stata using an R source file. Statistical Software Components S456847, Department of Economics, Boston College. https://ideas.repec.org/c/boc/bocode/s456847.html.

Neyman, J. 1934. On the two different aspects of the representative method: The method of stratified sampling and the method of purposive selection. *Journal of the Royal Statistical Society* 97: 558–625.

Nordberg, L. 1989. Generalized linear modeling of sample survey data. *Journal of Official Statistics* 5: 223–239.

Pacifico, D. 2014. sreweight: A Stata command to reweight survey data to external totals. *Stata Journal* 14: 4–21.

Pfeffermann, D. 1993. The role of sampling weights when modeling survey data. *International Statistical Review* 61: 317–337.

Pfeffermann, D., C. J. Skinner, D. J. Holmes, H. Goldstein, and J. Rasbash. 1998. Weighting for unequal selection probabilities in multilevel Models. *Journal of the Royal Statistical Society, Series B* 60: 23–40.

Pfeffermann, D., and M. Sverchkov. 2009. Inference under informative sampling. In Vol. 29B of *Handbook of Statistics: Sample Surveys: Inference and Analysis*, ed. D. Pfeffermann and C. R. Rao, chap. 39, 455–487. Amsterdam: Elsevier.

Potter, F. J. 1988. Survey of procedures to control extreme sampling weights. In *Proceedings of American Statistical Association, Survey Research Methods Section*, 453–458. American Statistical Association.

Rao, J. N. K., and C. F. J. Wu. 1985. Inference from stratified samples: Second-order analysis of three methods for nonlinear statistics. *Journal of the American Statistical Association* 80: 620–630.

———. 1988. Resampling inference with complex survey data. *Journal of the American Statistical Association* 83: 231–241.

Rivers, D. 2007. Sampling for web surveys. In *Proceedings of the 2007 Joint Statistical Meetings*. Salt Lake City, UT.
https://s3.amazonaws.com/yg-public/Scientific/Sample+Matching_JSM.pdf.

Robbins, M. W., B. Ghosh-Dastidar, and R. Ramchand. Forthcoming. Blending of probability and convenience samples as applied to survey of military caregivers. *Annals of Applied Statistics*.

Rosenbaum, P. R., and D. B. Rubin. 1983. The central role of the propensity score in observational studies for causal effects. *Biometrika* 70: 41–55.

Rubin, D. B. 1976. Inference and missing data. *Biometrika* 63: 581–592.

———. 1979. Using multivariate matched sampling and regression adjustment to control bias in observational studies. *Journal of the American Statistical Association* 74: 318–328.

———. 1983. Comment: Probabilities of selection and their role for Bayesian modeling in sample surveys. *Journal of the American Statistical Association* 78: 803–805.

Rust, K. 1985. Variance estimation for complex estimators in sample surveys. *Journal of Official Statistics* 1: 381–397.

Rust, K., and G. Kalton. 1987. Strategies for collapsing strata for variance estimation. *Journal of Official Statistics* 3: 69–81.

Särndal, C.-E. 2007. The calibration approach in survey theory and practice. *Survey Methodology* 33(2): 99–119.

Särndal, C.-E., B. Swensson, and J. Wretman. 1992. *Model Assisted Survey Sampling*. New York: Springer.

Schonlau, M. 2005. Boosted regression (boosting): An introductory tutorial and a Stata plugin. *Stata Journal* 5: 330–354.

Schonlau, M., A. van Soest, and A. Kapteyn. 2007. Are "webographic" or attitudinal questions useful for adjusting estimates from web surveys using propensity scoring? *Survey Research Methods* 1: 155–163.

Schonlau, M., K. Zapert, L. P. Simon, K. H. Sanstad, S. M. Marcus, J. Adams, M. Spranca, H. Kan, R. Turner, and S. H. Berry. 2004. A comparison between responses from a propensity-weighted web survey and an identical RDD survey. *Social Science Computer Review* 22: 128–138.

Simon, H. A. 1956. Rational choice and the structure of the environment. *Psychological Review* 63: 129–138.

Sirken, M. G. 1970. Household surveys with multiplicity. *Journal of the American Statistical Association* 65: 257–266.

Smith, T. M. F. 1976. The foundations of survey sampling: A review. *Journal of the Royal Statistical Society, Series A* 139: 183–204.

Spencer, B. D. 2000. An approximate design effect for unequal weighting when measurements may correlate with selection probabilities. *Survey Methodology* 26: 137–138.

Squiers, L. B., D. Brown, S. Parvanta, S. Dolina, B. Kelly, J. A. Dever, B. G. Southwell, A. Sanders, and E. M. Augustson. 2016. The SmokefreeTXT (SFTXT) Study: Web and mobile data collection to evaluate smoking cessation for young adults. *JMIR Research Protocols* 5(2): e134. http://www.researchprotocols.org/2016/2/e134/.

Squire, P. 1988. Why the 1936 literary digest poll failed. *Public Opinion Quarterly* 52: 125–133.

Sribney, W. undated. What does "completely determined" mean in my logistic regression output? http://www.stata.com/support/faqs/statistics/completely-determined-in-logistic-regression/.

SSI. 2015. *HLM - Hierarchical Linear and Nonlinear Modeling (HLM), version 7*. Skokie, IL: Scientific Software International. http://www.ssicentral.com/hlm/.

StataCorp. 2017. *Stata 15 Survey Data Reference Manual*. College Station, TX: Stata Press.

Statistics Canada. 2005. *BootVar Users Guide, Appendix C*.

Strobl, C., A.-L. Boulesteix, T. Kneib, T. Augustin, and A. Zeileis. 2008. Conditional variable importance for random forests. *BMC Bioinformatics* 9(307). http://www.biomedcentral.com/1471-2105/9/307.

Strobl, C., A.-L. Boulesteix, A. Zeileis, and T. Hothorn. 2007. Bias in random forest variable importance measures: Illustrations, sources and a solution. *BMC Bioinformatics* 8(25). http://www.biomedcentral.com/1471-2105/8/25.

Stuart, E. A., S. R. Cole, C. P. Bradshaw, and P. J. Leaf. 2011. The use of propensity scores to assess the generalizability of results from randomized trials. *Journal of the Royal Statistical Society, Series A* 174: 369–386.

Sturgis, P., N. Baker, M. Callegaro, S. Fisher, J. Green, W. Jennings, J. Kuha, B. Lauderdale, and P. Smith. 2016. Report of the inquiry into the 2015 British general election opinion polls. http://eprints.ncrm.ac.uk/3789/1/Report_final_revised.pdf. [Accessed 27 Sep 2017].

Therneau, T., B. Atkinson, and B. Ripley. 2017. *rpart: Recursive partitioning and regression trees*. R package version 4.1-11. http://CRAN.R-project.org/package=rpart.

Thompson, S. K. 2012. *Sampling*. 3rd ed. Hoboken, NJ: Wiley.

Thompson, S. K., and G. A. F. Seber. 1996. *Adaptive Sampling*. New York: Wiley.

Tillé, Y., and A. Matei. 2016. *sampling: Survey sampling*. R package version 2.8. http://CRAN.R-project.org/package=sampling.

Tourangeau, R., F. G. Conrad, and M. P. Couper. 2013. *The Science of Web Surveys*. New York: Oxford University Press.

U.S. Census Bureau. 2013. Source and Accuracy of Estimates for Income and Poverty in the United States: 2013 and Health Insurance Coverage in the United States: 2013. http://www2.census.gov/library/publications/2014/demo/p60-250sa.pdf.

―――. 2014. Source and Accuracy of Estimates for Income and Poverty in the United States: 2014 and Health Insurance Coverage in the United States: 2014. http://www2.census.gov/library/publications/2015/demo/p60-252sa.pdf.

Valliant, R. 2004. The effect of multiple weighting steps on variance estimation. *Journal of Official Statistics* 20: 1–18.

Valliant, R., J. M. Brick, and J. A. Dever. 2008. Weight adjustments for the grouped jackknife variance estimator. *Journal of Official Statistics* 24: 469–488.

Valliant, R., and J. A. Dever. 2011. Estimating propensity adjustments for volunteer web surveys. *Sociological Methods and Research* 40: 105–137.

Valliant, R., J. A. Dever, and F. Kreuter. 2013. *Practical Tools for Designing and Weighting Sample Surveys*. New York: Springer.

―――. 2016. *PracTools: Tools for designing and weighting survey samples*. R package version 0.4. http://CRAN.R-project.org/package=PracTools.

Valliant, R., A. H. Dorfman, and R. M. Royall. 2000. *Finite Population Sampling and Inference: A Prediction Approach*. New York: Wiley.

Wang, W., D. Rothschild, S. Goel, and A. Gelman. 2015. Forecasting elections with non-representative polls. *International Journal of Forecasting* 31: 980–991.

Westat. 2007. *WesVar 4.3 User's Guide*. Rockville, MD: Westat.

Willems, P., R. van Ossenbruggen, and T. W. E. Vonk. 2006. The Effects of Panel Recruitment and Management on Research Results. http://www.onderzoekpaleis.nl/ Diverse documenten/NOPVO2006-Barcelona-28112006.ppt.

Winter, N. 2002. svr: Stata module to compute estimates with survey replication (SVR) based standard errors. Statistical Software Components 427502, Department of Economics, Boston College. https://ideas.repec.org/c/boc/bocode/s427502.html.

Wolter, K. M. 2007. *Introduction to Variance Estimation*. 2nd ed. New York: Springer.

Wolter, K. M., X. Tao, R. Montgomery, and P. J. Smith. 2015. Optimal allocation for a dual-frame telephone survey. *Survey Methodology* 41: 389–401.

Zhou, H., M. R. Elliott, and T. E. Raghunathan. 2016a. Synthetic multiple-imputation procedure for multistage complex samples. *Journal of Official Statistics* 32: 231–256.

————. 2016b. Multiple imputation in two-stage cluster samples using the weighted finite population Bayesian bootstrap. *Journal of Survey Statistics and Methodology* 4: 139–170.

————. 2016c. A two-step semiparametric method to accommodate sampling weights in multiple imputation. *Biometrics* 72: 242–252.

# Author index

# Subject index